Elements of Foundation Design

Elements of Foundation Design

G. N. Smith
Senior Lecturer in Civil Engineering,
Heriot-Watt University, Edinburgh

E. L. Pole
Lecturer in Civil Engineering,
Heriot-Watt University, Edinburgh

GRANADA
London Toronto Sydney New York

Granada Publishing Limited–Technical Books Division
Frogmore, St Albans, Herts AL2 2NF
and
3 Upper James Street, London W1R 4BP
Suite 405, 4th Floor, 866 United Nations Plaza, New York, NY 10017, USA
117 York Street, Sydney, NSW 2000, Australia
100 Skyway Avenue, Rexdale, Ontario M9W 3A6, Canada
PO Box 84165, Greenside, 2034 Johannesburg, South Africa
61 Beach Road, Auckland, New Zealand

Copyright © 1980 by G. N. Smith and E. L. Pole

First published in Great Britain 1980 by Granada Publishing Limited

British Library Cataloguing in Publication Data
Smith, Geoffrey Nesbitt
Elements of foundation design.
1. Foundations
I. Title   II. Pole, E L
624'.15   TA775

ISBN 0-246-11429-0
ISBN 0-246-11215-8 Pbk

Phototypesetting by Parkway Group, London and Abingdon
Printed in Great Britain by Richard Clay (The Chaucer Press) Ltd,
Bungay, Suffolk

All rights reserved. No part of this publication may be reproduced, stored in a retrieval system, or transmitted in any form or by any means, electronic, mechanical, photocopying, recording or otherwise, without the prior permission of the publishers.

Granada ®
Granada Publishing ®

Contents

Preface — vii
Notation index — ix
1 Bearing capacity and settlement of foundations — 1
2 Piling — 55
3 Finite difference numerical techniques for foundations — 99
4 Reinforced earth — 141
5 Statistics and soil mechanics — 179
Appendix I Elements of matrix algebra — 205
Appendix II Flexibility and rigidity — 213
Appendix III The finite difference method — 215
Index — 221

Preface

Soil mechanics is a relatively young subject with a rate of development still vigorous enough to present an almost impossible task to the busy practising engineer wishing to keep abreast of the subject.

This book is not meant to be yet another elementary text on soil mechanics but is intended to present the reader with a description of recent developments, mainly in the field of foundation engineering. Accordingly our policy has been to avoid subjects such as earth pressure theory, sheet piling, etc., already adequately covered in soils text books and to concentrate on topics that they do not cover.

Chapters 1 and 2 contain practical, up-to-date methods, for the solution of foundation and piling problems. Chapter 3 illustrates how, with the aid of a programmable calculator, it is possible to analyse complex foundation problems. Chapter 4 deals with reinforced earth, now important enough to merit its own Department of Transport Memorandum[1]. Chapter 5 introduces the application of statistics to geotechnical situations. This subject is becoming of increasing importance to the soils engineer.

To derive full benefit from the book the reader should have a basic knowledge of soil mechanics, of a standard comparable to an undergraduate text book such as *Elements of Soil Mechanics*[2].

We both believe in the value of the worked example as a teaching aid and hope that the examples included in the text will enable an understanding of the subjects dealt with to be rapidly achieved and that the book will appeal to both student and practising engineer.

Our particular thanks must go to Mr R. T. Murray, of the Transport and Road Research Laboratory, Crowthorne and to Mr L. V. Leech, who both read through the manuscript and made many valuable suggestions. We would also like to acknowledge the encouragement given by Mr J. B. Boden, of the Building Research Establishment, Garston, in the preparation of material for chapters 4 and 5.

G. N. Smith
E. L. Pole

1. *Reinforced earth retaining walls and bridge abutments for embankments* (1978) Department of Transport Technical Memorandum (Bridges) BE3/78.
2. Smith, G. N. (1978) *Elements of Soil Mechanics for Civil and Mining Engineers.* 4th edn. London: Crosby Lockwood Staples.

Notation Index

The following is a list of the more important symbols used in the text.

A	Area, pore pressure coefficient
A_b	Area of base of pile, cross sectional area of pile
A_r	Cross sectional area of reinforcement
A_s	Area of surface of embedded length of pile shaft
B	Width, diameter, pore pressure coefficient
B'	Effective (or equivalent) width
C	Constant, correction factor
C_r	Static cone resistance
D_b	Depth of pile penetration into bearing stratum
D_c	Critical depth
D_r	Relative density
E	Modulus of elasticity
F	Factor of safety
H	Thickness, height
I	Index, moment of inertia, influence factor
K_a	Coefficient of active earth pressure
K_0	Coefficient of earth pressure at rest
K_p	Coefficient of passive earth pressure
K_s, K_u	Coefficient of lateral earth pressure
L	Length
L_a	Length of reinforcement in resistant zone
L'	Effective (or equivalent) length
M	Moment
N	Number, coefficient
N'	S.P.T. value corrected for overburden pressure
N_c, N_q, N_γ	Bearing capacity coefficients
P	Force
P_H	Horizontal applied force
P_V	Vertical applied force
Q	Subgrade reactive force

x

Q_b	Ultimate resistance of pile base
Q_s	Ultimate resistance of pile shaft
Q_u	Ultimate resistance of pile
R	Radius, reaction, resultant
R_t	Tensile strength of single reinforcing element
S	Settlement, sum
S_H	Horizontal spacing
S_v	Vertical spacing
S_L	Vertical line load
W	Weight, width
$\overline{X}$	True mean of a population
Z	Section modulus
a	Area
b	Width, breadth of a reinforcement strip
c, c_u	Unit cohesion with respect to total stress, undrained shear strength
c_a	Unit adhesion between pile and soil
c_r	Unit adhesion available along length of reinforcing element
c_v	Coefficient of consolidation
$\bar{c}_u$	Average undrained shear strength
c'	Unit cohesion with respect to effective stress
e	Eccentricity, void ratio
f_b	Unit penetration resistance of soil
f_ℓ	Limiting value of f_b
f_s	Ultimate unit skin friction for piles
g	Gravitational acceleration
i	Inclination factor
k_h	Coefficient of horizontal subgrade reaction
k_s	Coefficient of vertical subgrade reaction
m	Coefficient
m_v	Coefficient of volume compressibility
n	Porosity, coefficient, number
n_h	Constant of horizontal subgrade reaction for piles in granular soil
p	Pressure
p_a	Active earth pressure
p_0	Earth pressure at rest, total overburden pressure
p'_0	Effective overburden pressure
$\overline{p}'_0$	Average effective overburden pressure

p_p	Passive earth pressure
q	Ultimate bearing capacity, subgrade reactive pressure
r	Radius, radial distance
s	Shape factor
w_s	Uniform vertical surcharge
x	Horizontal distance
$\bar{x}$	Arithmetic mean of a sample
y	Vertical, horizontal distance, deflection
z	Vertical distance, depth
α	Angle, adhesion factor for piles
β	Angle, factor
γ	Unit weight
γ'	Submerged unit weight of soil
δ	Angle of friction between soil and foundation, wall or pile, deformation
ϵ	Strain, efficiency of pile group
θ	Angle
λ	Factor
μ	Poisson's ratio, coefficient of friction between soil and reinforcing element, consolidation factor
ρ	Settlement, density
σ	Normal stress, standard deviation
σ'	Effective normal stress
τ	Shear stress
ϕ	Angle of friction of soil, generally with respect to total stress
ϕ'	Angle of friction of soil with respect to effective stress
ψ	Angle

1. Bearing Capacity and Settlement of Foundations

Introduction
A foundation is the means by which the loads from a structure are transmitted into the geological deposits beneath it. These loads must be transmitted in such a manner that two criteria are satisfied:
  (i) There is no risk of shear failure in the underlying deposits.
  (ii) The resulting settlement of the foundation is acceptable to both the appearance and the function of the structure.

Bearing capacity terms
Shallow foundation
A foundation whose depth below the surface, z, is equal to, or less than, its least dimension, B.

Deep foundation
A foundation whose depth below the surface is greater than its least dimension.

Overburden pressure
The vertical pressure at a point, caused solely by the weight of the geological deposits above it. The effects of a foundation load are not included.

Ultimate bearing capacity
The average contact pressure between the foundation and the soil which will produce shear failure in the soil.

Maximum safe bearing capacity
The maximum contact pressure to which the soil can be subjected without risk of shear failure. This is based solely on the strength of the soil and is simply the ultimate bearing capacity divided by a suitable factor of safety.

Allowable bearing pressure
The maximum allowable net loading intensity on the soil allowing for both shear and settlement effects.

2 Elements of Foundation Design

Types of foundations

Strip foundation
A foundation of continuous length. The term is also used to describe a foundation whose length is considerably greater than its width.

Pad foundation
A spread footing designed to carry either one or more column loads. When underlying soil conditions permit, it is usual to provide each column of a structure with its own pad foundation but there are often occasions when a pad foundation carries two or more columns.

Raft foundation
This is a rather loose term used to denote any foundation that covers a large area. A raft foundation can vary from a fascine mattress carrying a road over a marsh to an extensive basement of deep beams and slabs supporting a large structure.

Piled foundation
Piling can be either end bearing, frictional or a combination of the two (see chapter 2).

Determination of the ultimate bearing capacity of a soil

Terzaghi's equations

These equations for the ultimate bearing capacity, q, of shallow strip, square and circular foundations are based on Prandtl's analysis of the effects of a punch penetrating into metal and were first published in 1943. They are well known and their derivation is described in most soils text books (Smith 1978).

For a strip footing:

$$q = cN_c + \gamma z N_q + 0.5\gamma B N_\gamma.$$

For a circular footing:

$$q = 1.3cN_c + \gamma z N_q + 0.3\gamma B N_\gamma \text{ (where B = diameter)}.$$

For a square footing:

$$q = 1.3cN_c + \gamma z N_q + 0.4\gamma B N_\gamma.$$

For a rectangular footing:

$$q = cN_c \left(1 + 0.3 \frac{B}{L}\right) + \gamma z N_q + 0.5\gamma B N_\gamma \left(1 - 0.2 \frac{B}{L}\right).$$

In the equations:
c = unit cohesion of the soil
γ = unit weight of the soil
z = depth from surface to underside of foundation
N_c, N_q and N_γ are bearing capacity coefficients which depend upon ϕ, the angle of friction of the soil, and can be obtained from fig. 1.1.

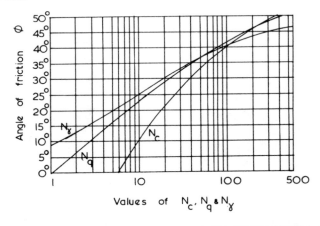

ϕ	0°	5°	10°	15°	20°	25°	30°	35°	40°	45°	50°
N_c	5.7	7.3	9.6	12.9	17.7	25.1	37.2	57.8	95.7	172	348
N_q	1.0	1.6	2.7	4.4	7.4	12.7	22.5	41.4	81.3	173	415
N_γ	0.0	0.5	1.2	2.5	5.0	9.7	19.7	42.4	100	298	1153

Fig. 1.1 Terzaghi's bearing capacity coefficients

Value of N_c for cohesive soils

From fig. 1.1 is seen that when $\phi = 0°$, $N_c = 5.7$, $N_q = 1.0$ and $N_\gamma = 0$. Hence, according to Terzaghi, when $\phi = 0°$; the ultimate bearing capacity of a strip footing is:

$$q = 5.7c + \gamma z$$
$$= 5.7c \text{ for a surface footing.}$$

4 Elements of Foundation Design

Skempton (1951) showed that for cohesive soils ($\phi = 0°$) N_c tends to increase with depth and suggested that values of N_c greater than 5.7 could be inserted into the Terzaghi equations when appropriate. His suggested values for N_c are shown in fig. 1.2.

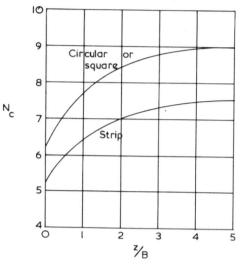

Fig. 1.2 Variation of coefficient N_c with depth (Skempton)

Net ultimate bearing capacity and safe bearing capacity

The foregoing equations give values for q, the ultimate bearing capacity of the soil on which the foundation is to sit. To obtain the safe bearing capacity the relevant value for q must be divided by a factor of safety against bearing capacity failure, generally taken as 3.

However, it must be remembered that, due to the excavation of material in order that the foundation can be installed at depth z, there will be a relief in vertical pressure at the foundation level of γz. This means that the factor of safety should be applied to the *net* and not to the *gross* value of ultimate bearing capacity. If the excavation is subsequently backfilled the overburden pressure is restored and the net bearing capacity therefore applies.

For a strip footing:

$$q_{net} = cN_c + \gamma z N_q + 0.5\gamma B N_\gamma - \gamma z$$
$$= cN_c + \gamma z(N_q - 1) + 0.5\gamma B N_\gamma$$

and safe bearing capacity = $(q_{net}/3) + \gamma z$.

When the footing is at the surface of the soil, $z = 0$ and safe bearing capacity = $q/3 = cN_c/3$.

Example 1.1

A circular foundation is to carry a concentric column load of 850 kN.

The foundation will be installed at a depth of 1.0 m in a partially saturated soil which has an angle of friction of 10°, a unit cohesion of 40 kN/m² and a unit weight of 18 kN/m³.

Ignoring the weight of the foundation and given that the excavated soil will not be backfilled, determine a suitable dimension for the diameter of the foundation. Assume a factor of safety against bearing capacity failure of 3.

For a circular foundation:

$$q_{net} = 1.3 cN_c + \gamma z(N_q - 1) + 0.3 \gamma B N_\gamma.$$

From fig. 1.1, for $\phi = 10°$, $N_c = 9.6$, $N_q = 2.7$ and $N_\gamma = 1.2$

$$\therefore q_{net} = 1.3 \times 9.6 \times 40 + 18 \times 1 \times 1.7 + 0.3 \times 18 \times B \times 1.2$$
$$= 529.8 + 6.48 B.$$

Safe bearing capacity:

$$= \frac{529.8 + 6.48B}{3} + 18 \times 1$$

$$= 194.6 + 2.16B.$$

Hence, ignoring the weight of the foundation:

$$\frac{850}{\pi B^2/4} = 194.6 + 2.16 B$$

or
$$B^3 + 90.1 B^2 - 501.1 = 0$$

whence
$$B = 2.328$$

so that the foundation requires a diameter of about 2.35 m.

Effective stress form of the equations

The foregoing Terzaghi equations are expressed in terms of total stress but are equally valid when expressed in terms of effective stress.

6 Elements of Foundation Design

For a strip footing:

$$q = c' N_c + p'_0 N_q + 0.5 \gamma' B N_\gamma$$

where c' = unit cohesion of soil with respect to effective stresses,
p'_0 = effective overburden pressure (total pressure–pore water pressure)
N_c, N_q and N_γ are bearing capacity coefficients which depend upon the value of ϕ', the angle of internal friction with respect to effective stress, and can be obtained from fig. 1.1.
γ' = submerged unit weight of soil.

Example 1.2

A square footing is required to carry a column load of 560 kN with a factor of safety against bearing capacity failure of 3. The foundation is to be placed at a depth of 1.5 m below the surface of a loose granular soil where the water table will be at foundation level.

The angle of friction of the soil can be assumed to be 28° and the bulk unit weights of the soil above and below the water table are 16.5 and 18.3 kN/m³ respectively.

If the excavation will be backfilled to its original level, determine a suitable size for the foundation.

For a square foundation:

$$q_{net} = 1.3 c N_c + p'_0 (N_q - 1) + 0.4 \gamma' B N_\gamma.$$

The soil is granular, so $c' = 0$ and the equation becomes:

$$q_{net} = p'_0 (N_q - 1) + 0.4 \gamma' B N_\gamma.$$

From fig. 1.1, for $\phi' = 28°$, $N_q \approx 20$ and $N_\gamma \approx 17$.

Then
$$q_{net} = 1.5 \times 16.5 \times 19 + 0.4 B \times 17 (18.3 - 9.81)$$
$$= 470.25 + 57.7B \text{ kN/m}^2.$$

Safe bearing capacity = $156.8 + 19.2B$ kN/m².

Hence, ignoring the weight of the foundation:

$$560/B^2 = 156.8 + 19.2B$$
or
$$560 = 156.8B^2 + 19.2B^3$$
whence
$$B = 1.72 \text{ m}.$$

Example 1.3
A block of flats is to be erected on a raft foundation at a site where the soil profile consists of 1 m of fill material overlying a deep bed of saturated clay. The dimensions of the raft will be 24 m × 12 m and it will be founded at a depth of 2 m below the surface of the ground.

A site investigation has shown that the undrained shear strength of the clay increases from 50 kN/m² at proposed foundation level to 80 kN/m² at a depth of 14 m. The unit weights of the fill and the clay are 17 and 20 kN/m³.

If the gross pressure at foundation level from the raft will be 175 kN/m², check the factor of safety against bearing capacity failure, which should not be less than 2.5.

No ground water conditions have been noted, ∴ use total stress equations.

For a rectangular foundation:

$$q_{net} = cN_c\left(1 + 0.3\frac{B}{L}\right) + \gamma z(N_q - 1) + 0.5\gamma BN_\gamma\left(1 - 0.2\frac{B}{L}\right).$$

With $\phi = 0°$ (soil is a saturated clay), $N_c = 5.7$, $N_q = 1.0$ and $N_\gamma = 0$

$$\therefore q_{net} = cN_c\left(1 + 0.3\frac{B}{L}\right).$$

The cohesion of the soil varies with depth. In this type of problem it is general practice to assume that the unit cohesion value to be used in the bearing capacity formula is the average value between foundation level and a depth of 2/3 B below it.

At 2/3 B below foundation level,

$$c_u = 50 + (80 - 50)\frac{8}{12} = 70 \text{ kN/m}^2$$

$$\therefore \text{average } c_u = \frac{1}{2}(50 + 70) = 60 \text{ kN/m}^2.$$

Hence, $q_{net} = 60 \times 5.7\left(1 + 0.3 \times \frac{12}{24}\right) = 393.3 \text{ kN/m}^2.$

8   *Elements of Foundation Design*

Let factor of safety = F.

Then safe bearing capacity:

$$= \frac{393.3}{F} + (1 \times 17 + 1 \times 20)$$

$$= \frac{393.3}{F} + 37.$$

Gross contact pressure from the raft:

$$= 175 \text{ kN/m}^2$$

$$\therefore F = \frac{393.3}{175 - 37} = 2.85 > 2.5. \ldots . \text{O.K.}$$

Modification of Terzaghi's equations for deep foundations

In the case of a deep foundation (z > B) the ultimate bearing capacity is increased due to the side friction generated around the surface of the excavation. This increase can be allowed for by the following technique.

Consider a rectangular foundation, of dimensions B × L, founded at a depth z.

Ultimate bearing capacity of foundation = $q + 2(L + B)zf_s$

where q = ultimate bearing capacity of a shallow foundation of dimensions B × L

f_s = unit skin friction (as calculated for the shaft of a pile—see chapter 2).

Choice of soil parameters

As the bearing capacity equations can be expressed in terms of either total or effective stress, the design engineer has to decide which conditions most closely fit his problem.

For free draining soils the effective stress parameters will apply in the soil at all stages of construction and life of the foundation.

When a foundation load is applied to a saturated clayey soil the excess pore water pressures generated can only dissipate with time.

Bearing capacity and settlement of foundations 9

Therefore, for most foundation problems of this type, the short term stability, i.e. the stability immediately after construction, should be checked in terms of total stress.

For the estimation of the long term stability of a foundation on clayey soil, i.e. after the dissipation of excess pore pressures, the effective stress parameters are relevant. However, it should be remembered that dissipation of pore water pressure is accompanied by consolidation of the soil which generally leads to an increase in strength.

It is therefore rarely necessary to check the long term stability of a foundation.

Further bearing capacity equations

Since Terzaghi's original equations were published several authors have proposed modifications and improvements.

Meyerhof (1951) evolved equations for deep foundations and Hansen (1957) proposed equations for cohesive soils which are considered to be more accurate than those of Terzaghi. Balla (1962) proposed a theory which is considered to be the most appropriate for granular soils, or soils possessing little cohesion.

The fact remains, however, that Terzaghi's original equations are still the most widely used and the reason is not far to seek. They can be used for all types of soils and give conservative values for the ultimate bearing capacity which often lead to safe bearing capacities close to the allowable bearing pressures when settlement effects are taken into account.

Determination of bearing capacity values for granular soils

Although Terzaghi's equations can be applied to all types of soil they are often impractical with granular soils because it is almost impossible to obtain undisturbed samples that are sufficiently satisfactory to be meaningfully tested in the laboratory.

However, with these soils, the most decisive factor in the determination of the allowable bearing pressure is the amount of settlement that the foundation will experience. Reasonably satisfactory predictions of settlement in granular soils can be made from the results of in situ tests and it is now general practice to estimate the required parameters on this basis.

10 *Elements of Foundation Design*

There are two main types of test, those that penetrate into the soil and those that are applied at its surface.

Penetration tests

The two most commonly used penetration tests are the Standard penetration test and the Dutch cone penetration test. Both are extremely simple to carry out, which accounts for their popularity.

(a) The Standard penetration test (S.P.T.)

The Standard penetration test, described in most soils text books, is a dynamic test carried out in a borehole and consists in determining the S.P.T value, N, i.e. the number of blows of a standard weight that are required to drive a 35 mm interior diameter sampling tube through a depth of 300 mm into the soil, having already driven it through a depth of 150 mm.

Terzaghi and Peck (1948) established an empirical relationship between the S.P.T. value, N, and the bearing pressure, p, that results in settlements not greater than 25 mm for foundations of varying widths. This relationship is illustrated in fig. 1.3 and is applicable to foundations unaffected by ground water. The authors suggested that, when the ground water level is at a depth below the foundation of less than B, the values of p quoted in the figure should be halved, as settlements are approximately doubled.

Terzaghi and Peck also suggested that in saturated (i.e. below the water table) fine and silty sands the N value could be altered due to the low permeability of the soil.

They suggested that N be corrected by using the formula:

$$N_{corrected} = 15 + \frac{N - 15}{2}$$

where N = actual blow count obtained from test.

Various criticisms and suggestions have been applied to Terzaghi and Peck's N-p relationship since its introduction.

Meyerhof (1956) considered that the values of p were conservative and could be increased by 50%. He also considered that, whilst

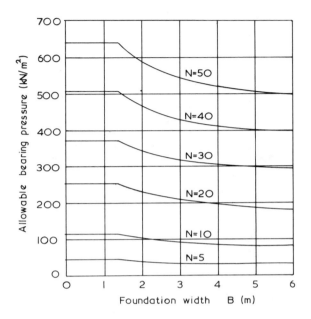

Fig. 1.3 Relationship between allowable bearing pressure, p, and S.P.T. value, N (Terzaghi and Peck, 1948)

'agreeing with the correction for submerged silty and fine sands, the practice of halving allowable pressure values for foundations subjected to ground water conditions was possibly overconservative in that the influence of the water table is already included in the measured value of N and therefore need not be allowed for again.

Correction for overburden pressure
The effect of the removal of the overburden pressure prior to the test has been studied by a number of investigators: Gibbs and Holtz (1957), Thorburn (1963), Peck and Bazaraa (1969), Peck *et al.* (1974).

The correction factor to be applied to the actual blow count, N, as proposed by Peck and Bazaraa, is reproduced in fig. 1.4 and is considered to give realistic values. It is seen that the correction imposes a significant increase in N for shallow depths.

12 Elements of Foundation Design

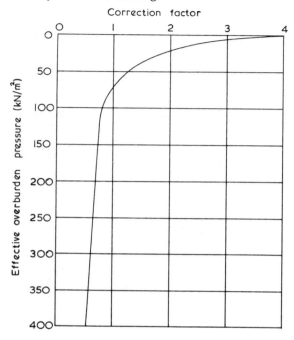

Fig. 1.4 Correction factor to be applied is the S.P.T. value, N, to allow for the effect of overburden pressure (after Peck and Bazaraa, 1969)

Peck and Bazaraa agreed with Meyerhof that Terzaghi and Peck's p values were low and they also recommended that they be increased by 50%. They further suggested that, when the ground water level is at a depth of less than B/2 below the foundation, the 25 mm settlement of the Terzaghi and Peck curve will be increased by a factor K, where K is the ratio of the effective overburden pressure at 0.5B below the foundation when the soil is unsubmerged to that at the same level when the soil is submerged.

Example 1.4
A total uniformly distributed load of 9.5 MN is to be supported by a raft founded at a depth of 1.5 m in a granular soil. The bulk unit weight of the soil is 18 kN/m^3 and ground water level occurs at a depth of 10 m below the surface. S.P.T. values, taken at 1 m and 6 m depths, were N = 8 and N = 28 respectively.

The proposed dimensions of the raft are 8 m × 6 m. Check that

Bearing capacity and settlement of foundations 13

the average settlement of the raft will not exceed 25 mm.

Effective overburden pressure at 1 m depth = $1 \times 18 = 18 \text{ kN/m}^2$.
From fig. 1.4, corrected $N = N' \approx 8 \times 2.4 = 19$.
Effective overburden pressure at 6 m depth = $6 \times 18 = 108 \text{ kN/m}^2$.
From fig. 1.4, corrected $N = N' \approx 28 \times 0.8 = 22$.
Average N' value ≈ 20.
From fig. 1.3, for $N' = 20$ and $B = 6$ m, allowable bearing pressure = 180 kN/m^2.

According to Peck and Bazaraa this can be increased by 50%. Thus, design allowable bearing pressure = $1.5 \times 180 = 270 \text{ kN/m}^2$.

∴ total foundation load to cause an average settlement of 25 mm

$$= \frac{270}{1000} \times 8 \times 6 = 12.96 \text{ MN}.$$

As the actual foundation load is 9.5 MN, the settlement of the raft will average less than 25 mm.

Actual average settlement = $\frac{9.5}{12.96} \times 25 = 18$ mm.

Example 1.5
What would the average settlement of the raft in example 1.4 have been if the ground water level had been at a depth of 1.5 m, all other conditions remaining the same?

As the ground water level is at 1.5 m depth, the effective stress at 1 m depth is unaltered.

$$\therefore N' = 19 \text{ (as before)}.$$

Effective stress at 6 m depth = $6 \times 18 - 9.81 \times 4.5 = 63.9 \text{ kN/m}^2$.

From fig. 1.4, $N' \approx 1.1 \times 28 = 31$.

$$\therefore \text{Average } N' = 25.$$

From fig. 1.3, and increasing by 50%,

Allowable bearing pressure = $1.5 \times 240 = 360 \text{ kN/m}^2$ for a 6 m wide raft.

This now has to be corrected for the high ground water conditions.

$$K = \frac{\text{Overburden pressure of unsubmerged soil at depth } 0.5B \text{ below foundation}}{\text{Effective overburden pressure at same depth with G.W.L. at 1.5 m depth}}$$

$$= \frac{18(1.5 + 3)}{18(1.5 + 3) - 9.8 \times 3} = 1.57.$$

This indicates that, if the foundation were stressed to a bearing pressure of 360 kN/m², the total settlement would be in the region of $25 \times 1.57 = 39$ mm.

$$\text{Actual stress from foundation load} = \frac{9.5 \times 1000}{6 \times 8} = 198 \text{ kN/m}^2.$$

$$\therefore \text{Actual settlement of raft} \approx \frac{198}{360} \times 39 = 22 \text{ mm}$$

which is less than the normally permitted maximum settlement of 25 mm. The proposed dimensions of the raft are therefore satisfactory.

(b) The Dutch cone penetration test

The apparatus consists of a cylindrical probe, of 1000 mm² cross sectional area, headed by a cone with an apex angle of 60°. The probe is forced down through the soil at a steady rate and the resistance to penetration measured. If desired the point resistance and the resistance to side friction can be measured separately. The main disadvantage of the test is that it is not always possible to penetrate an underlying hard layer of soil.

The results of the test, as with the Standard penetration test, can be used to predict the possible settlement of a foundation placed on the soil tested.

The method of prediction generally used is that proposed by de Beer and Martens (1957):

$$\text{Constant of compressibility}, C_s = 1.5 \frac{C_r}{p_{0_1}}$$

where C_r = static cone resistance (kN/m²)
p_{0_1} = effective overburden pressure at point tested.

Total immediate settlement, $\rho_i = \dfrac{H}{C_s} \log_e \dfrac{p_{0_2} + \Delta \sigma_z}{p_{0_2}}$

where $\Delta \sigma_z$ = vertical stress increase at the centre of the consolidating layer of thickness H.

p_{0_2} = effective overburden pressure at the centre of the layer before any excavation or load application.

After a series of observations on actual structures it is now generally felt that de Beer and Martens' method can give predicted settlements that are high, sometimes twice as much as the actual measured value. In view of this, the relationship between C_s and C_r proposed by Meyerhof (1965),

$$C_s = 1.9 \dfrac{C_r}{p_{0_1}}$$

is often preferred. This tends to give more conservative estimates of settlement.

A further method by which Dutch cone test results can be used to predict settlement has been prepared by Schmertmann (1970). It is based on the fact that the greatest strain beneath a foundation on sand is at a depth of approximately B/2 beneath it and on the assumption that significant vertical stresses do not extend beyond a depth of 2B.

The method is considered by many to give more realistic predictions than de Beer and Martens and has the advantage that it can predict settlement not only immediately after construction but also after several years as it can allow for creep effects.

Schmertmann's formula for the settlement of a foundation on sand can be expressed as:

$$\rho = C_1 C_2 \Delta \sigma_z \sum_0^{2B} \dfrac{I_z \Delta z}{2C_r}$$

where $\Delta \sigma_z$ = increase in vertical stress at depth z due to foundation load

C_r = static cone resistance
C_1 = correction factor for depth
C_2 = correction factor for creep
I_z = strain influence factor
Δz = thickness of layer under consideration.

With the knowledge that the greatest strain occurs at about B/2,

16 Elements of Foundation Design

the variation of I_z with depth is as shown in fig. 1.5.

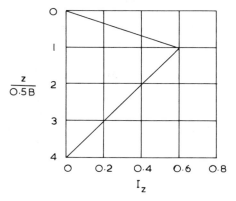

Fig. 1.5 Variation of I_z with z (Schmertmann, 1970)

C_1 and C_2 are obtained from the following formulae:

$$C_1 = 1 - 0.5 \left(\frac{p'_0}{\Delta\sigma_z}\right)$$

where p'_0 = initial effective overburden pressure at the foundation

$$C_2 = 1 + 0.2 \log_{10}\left(\frac{t}{0.1}\right)$$

where t = time in years.

The operation of both methods is best shown by means of an example.

Example 1.6

A 20 m × 4 m spread footing is to be founded in a deep sand layer of unit weight 18 kN/m³ at a depth of 1.8 m below its surface (fig. 1.6A).

When the full structure is completed, estimated construction time = 1 year, the gross loading from the foundation will be a uniform pressure of 204 kN/m².

Dutch cone tests carried down through the sand gave the results shown in fig. 1.6B.

Determine: (i) Settlement immediately after construction.
(ii) Settlement 4 years after construction.

Bearing capacity and settlement of foundations 17

For both methods the scatter of the test results must be approximated to an idealised form suitable for use in the calculations (fig. 1.6C).

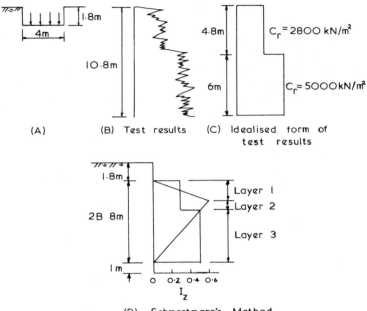

Fig. 1.6

(a) Schmertmann's method

The procedure is to superimpose fig. 1.5 on to fig. 1.6C and to divide the resulting diagram into slices that are convenient for analysis (fig. 1.6D). It is seen that, for this example, the minimum number of slices is three.

The calculations are best set out as in table 1.1:

Table 1.1

Layer no.	Thickness Δz (mm)	C_r (kN/m^2)	I_z (average)	$\dfrac{I_z \Delta z}{2C_r}$
1	2000	2800	0.3	0.1071
2	1000	2800	0.55	0.0982
3	5000	5000	0.25	0.125
				Σ 0.3303

18 Elements of Foundation Design

$$C_1 = 1 - 0.5 \left(\frac{1.8 \times 18}{204 - 32.4}\right) = 0.9056.$$

Note. Relief in vertical pressure due to excavation for foundation = $1.8 \times 18 = 32.4 \text{ kN/m}^2$.

(i) Settlement immediately after contruction.

The construction time will be one year and a simple assumption, so that C_2 can be evaluated, is that, instead of being applied gradually over one year, the foundation load is applied suddenly after six months, i.e. t = 0.5.

Then $\qquad C_2 = 1 + 0.2 \log_{10} \left(\dfrac{0.5}{0.1}\right) = 1.14.$

∴ Settlement at end of construction:

$$= 0.9056 \times 1.14 \times (204 - 32.4) \times 0.3303$$
$$= 59 \text{ mm}.$$

(ii) Settlement 4 years after construction:

$$t = 4.5 \quad \therefore C_2 = 1.33$$

and $\qquad \rho = 0.9056 \times 1.33 \times 171.6 \times 0.3303 = 68 \text{ mm}.$

(b) de Beer and Martens' method

The idealised form of the soil conditions (fig. 1.6C) is used with this method. For simplicity, let us divide the soil beneath the foundation into three equal strips, each 3 m in thickness. Again, the calculations are best tabulated (tables 1.2 and 1.3).

Table 1.2

Slice no.	Depth below foundation to centre of layer, z (m)	p_{0_2} (kN/m²)
1	1.5	$3.3 \times 18 = 59.4$
2	4.5	$6.3 \times 18 = 113.4$
3	7.5	$9.3 \times 18 = 167.4$

Hence, according to this method, the settlement of the foundation immediately after construction, and then for all time = 69 mm.

Determination of $\Delta\sigma_z$ increments

The simplest, and quite often accurate enough, method for the

Bearing capacity and settlement of foundations

Table 1.3

Slice No.	p_{0_2} (kN/m²)	C_r (kN/m²)	$C_s = \dfrac{1.9 C_r}{p_{0_2}}$ (Meyerhof)	Gross $\Delta\sigma_z$ (kN/m²)	Net $\Delta\sigma_z$ (kN/m²)	$\log_e \dfrac{p_{0_2} + \Delta\sigma_{z\,net}}{p_{0_2}}$ (A)	$\dfrac{H}{C_s} \times$ (A)
1	59.4	2800	90	188	156	1.288	42.9
2	113.4	5000	84	103	71	0.486	17.4
3	167.4	5000	57	63	31	0.170	8.9
							Σ 69.2

determination of the increase in vertical stress at a depth due to a foundation load is to assume that this load spreads itself downwards throughout the soil at a slope of 1 horizontal to 2 vertical. The reader might like to check that, with this assumption, the gross values of $\Delta\sigma_z$ are:

Centre of slice 1 = 138 kN/m²
Centre of slice 2 = 78 kN/m²
Centre of slice 3 = 52 kN/m².

A more sophisticated technique has been evolved by Steinbrenner (1934) and is described by Smith (1978). The values of $\Delta\sigma_z$ obtained by this method have been used in the preceding example.

Relationship between C_r and N

Various workers have attempted to arrive at a formula which would relate the S.P.T. value, N, to its corresponding static cone resistance, C_r. Results have been disappointing but a formula generally in use, although inaccurate, is:

$$C_r = 400N \text{ (kN/m}^2\text{)}$$

where N = the uncorrected S.P.T. value.

To emphasise the shortcomings of the formula the reader's attention is drawn to the work of Meigh and Nixon (1961) who showed that, over a number of sites, C_r values varied from 430N to 1930N (kN/m²).

The pressuremeter

Both the Standard penetration test and the Dutch cone penetration test have the advantages of cheapness and simplicity of operation.

20 Elements of Foundation Design

The Dutch cone test is particularly useful in that it can provide a continuous profile of soil penetration resistance to depth.

However, the interpretation of the results from both tests involves the use of empirical rules which may, or may not, lead to accurate results.

It has been realised for a long time that, for relatively homogeneous soil conditions, the only in situ test that gives results that can be analysed to yield reasonably accurate estimations of soil parameters is the plate loading test. This test, briefly described in the next section, has the disadvantages of being both costly and time consuming.

A recent development in in situ testing techniques is the pressuremeter which can provide a quicker and cheaper form of test.

Originally developed by Menard (1957) the pressuremeter consists of an inflatable rubber bag which is lowered down a borehole in the soil to be tested. The bag is inflated with nitrogen against the sides of the borehole and, by measuring gas pressure and strain effects, it is possible to determine such soil properties as elastic modulus and undrained shear strength.

The main disadvantage of the Menard pressuremeter is that it can only be used down a borehole and the quality of the results obtained is heavily dependent upon the degree of disturbance suffered by the soil forming its sides.

Since its inception the pressuremeter has been subjected to modification by various workers but a major improvement has resulted from work carried out by Baguelin at the Laboratoire des Ponts et Chausées, Paris (1974) and by Wroth at Cambridge University (1975). They have modified the apparatus to make it self boring. Two models of the self boring pressuremeter exist, one on each side of the Channel, but the basic operation of both is very similar.

It is claimed that the self boring pressuremeter has the advantages that the soil into which it is introduced suffers an almost negligible amount of disturbance. It is possible to obtain several soil parameters, such as undrained shear strength, Young's Modulus and the lateral earth pressure coefficient, K_0, all from the use of proven theories rather than from empirical rules.

The disadvantages appear to be that the apparatus is fairly complex, requires specialist operators and, until its durability has been proved, cannot be considered as being very robust. It is also not possible for the instrument to penetrate hard layers such as dense gravel, mudstone, etc.

The plate loading test

This is a semi-direct means for the determination of the bearing capacity of a relatively homogeneous soil and, whilst generally used for granular soils, can be used on clays.

The soil is excavated down to the proposed foundation level and a model footing, made up from thick plates of steel, is subjected to an increasing vertical stress. The settlement at various steps in the loading is noted.

The test plate is generally 305 mm square but round plates, and larger sizes, are sometimes used. Typical test arrangements and the results obtained are illustrated in fig. 1.7.

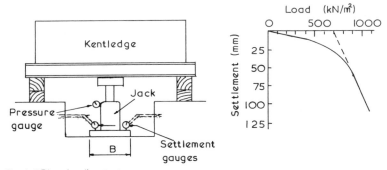

Fig. 1.7 Plate loading test

The ultimate bearing capacity of the soil is taken either as the loading pressure at which the load/settlement relationship approaches the vertical or that which produces a settlement of 25 mm.

With granular soils the usual problem is to determine the allowable bearing pressure for a particular foundation and it is generally taken to be the pressure that will cause an average settlement of the foundation of 25 mm.

The settlement of a square footing, kept at a constant pressure, increases as the size of the footing increases.

Terzaghi and Peck (1948) produced the following relationship:

$$S = S_1 \left(\frac{2B}{B + 0.3} \right)^2$$

where S_1 = settlement of a loaded area of 0.305 m × 0.305 m under a given loading intensity of p
S = settlement of a square, or rectangular, footing of width B m under the same pressure p.

Knowing the size of the footing, B and the fact that $S = 25$ mm, it is possible to work backwards and obtain S_1. The test pressure corresponding to S_1 is the allowable bearing pressure. The process is one of iteration, starting with an assumed value for B.

It should be noted that the zone of soil affected by a plate load test is much smaller than will be affected beneath the full foundation. Horizontal and vertical variations in the soil may not be reflected in the plate load test results and it is useful to have borings taken at the load test site to check that the soil is sufficiently homogeneous for plate load tests to be realistic.

Procedure in foundation design

The bearing capacity of a proposed foundation can be obtained either by inserting measured soil parameters into the relevant bearing capacity formula or from the results of in situ tests.

When no information is available at the start of a design project reference can always be made to tables of safe bearing capacities prepared by various authorities.

Table 1.4 lists typical safe bearing capacities (based on CP 2004 'Foundations') that can be associated with particular soil types.

With the use of such values a preliminary estimate of the foundation size can be made. At this stage the foundation depth should be decided upon from a consideration of such factors as the depth of frost penetration, position of the water table and the location of a suitable soil bearing stratum. Afterwards the size of the foundation can be checked in terms of soil parameters or in situ test results, as described previously.

Table 1.4 Values of safe bearing capacity (kN/m^2)

Soil type	Dry	Submerged
Compact well graded sands and gravel-sand mixtures	500	250
Loose well graded sands and gravel-sand mixtures	300	150
Compact uniform sands	300	150
Loose uniform sands	160	80
Very stiff boulder clays and hard clays with a shaly structure		500
Stiff clays and sandy clays		300
Soft clays and silts		160
Very soft clays and silts		25

Bearing capacity and settlement of foundations 23

Settlement of foundations

Note. This subject has been fairly well covered in chapters 9 and 10 of the companion volume to this text, *Elements of Soil Mechanics for Civil and Mining Engineers*, and only a brief summary will be presented here.

(a) *Cohesive soils*
(i) Immediate settlement
An expression for the *mean* settlement of a reinforced concrete surface foundation was suggested by Skempton in 1951.

$$\rho_i = \frac{pB(1 - \mu^2)I_p}{E}$$

where p = uniform contact pressure
B = width of foundation
E = modulus of elasticity of soil
μ = Poisson's ratio of soil
I_p = influence factor depending upon the dimensions of the foundation, i.e. length, L and width, B.

The values for I_p suggested by Skempton are given in table 1.5.

Table 1.5

L/B	I_p
circle	0.73
1 (square)	0.82
2	1.00
5	1.22
10	1.26

Typical values for μ are:

Saturated clay	0.4–0.5
Partially saturated clay	0.1–0.3
Silt	0.3–0.35
Sand	0.2–0.4

The above formula, based on the theory of elasticity, applies to a footing supported on a semi-infinite soil, i.e. in practical terms, a soil layer of not less than 2B (ideally, not less than 4B) in thickness. Janbu et al. (1956) proposed a formula for the mean settlement, ρ_i, when the consolidating layer is of a finite thickness H:

$$\rho_i = \mu_0 \mu_1 \, pB \text{ (for saturated clays)}.$$

Values of the coefficicients μ_0 and μ_1 are dependent upon the ratio H/B and L/B and are presented in fig. 1.8.

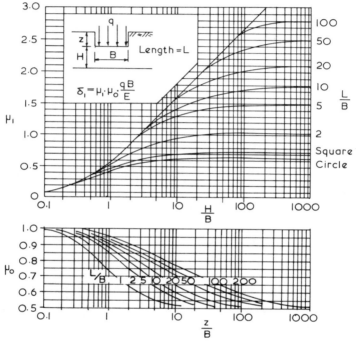

Fig. 1.8 Coefficients μ_0 and μ_1 for the determination of the mean immediate settlement of a foundation (after Janbu et al., 1956)

When the soil deposits beneath a foundation consist of a series of layers the mean settlement of the foundation can be obtained from the principle of superposition, as shown in the following example.

Example 1.7

A 7 m thick layer of clay has a value of $E = 25 \, MN/m^2$ and overlies a further 4 m thick layer of clay which has $E = 40 \, MN/m^2$. The lower

Bearing capacity and settlement of foundations 25

clay overlies bedrock which can be regarded as incompressible.

A square foundation slab of dimensions 5 m × 5 m is to be constructed at a depth of 2 m below the surface of the upper clay and will impose a uniform bearing pressure of 85 kN/m².

Estimate the value of mean immediate settlement of the foundation.

The problem is illustrated in fig. 1.9.

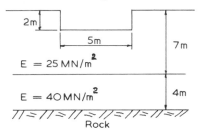

Fig. 1.9 Example 1.7

Consider upper layer of clay:

$z/B = 2/5 = 0.4$; $L/B = 5/5 = 1.0$ and $H/B = 5/5 = 1.0$.

From fig. 1.8, $\mu_0 = 0.9$ and $\mu_1 = 0.45$

$$\therefore \rho_i = 0.9 \times 0.45 \times \frac{85 \times 5}{25} = 6.9 \text{ mm}.$$

This is not the total settlement of the foundation as there will be some compression in the lower clay. This can be found by the following procedure:

If this lower clay had extended from the ground surface for the full depth of 11 m, then $z/B = 0.4$; $L/B = 1.0$ and $H/B = 9/5 = 1.8$, which gives $\mu_0 = 0.9$ and $\mu_1 = 0.55$.

Hence ρ_i would have been $0.9 \times 0.55 \times 85 \times \frac{5}{40} = 5.3$ mm.

If the clay had extended from the ground surface to the 7 m depth, then $z/B = 0.4$; $L/B = 1.0$ and $H/B = 5/5 = 1.0$ which gives $\mu_0 = 0.9$ and $\mu_1 = 0.45$ and ρ_i would have been

$$0.9 \times 0.45 \times 85 \times \frac{5}{40} = 4.3 \text{ mm}.$$

Hence ρ_i due to compression of lower clay $= 5.3 - 4.3 = 1.0$ mm.

∴ Total mean immediate settlement of the foundation $= 6.9 + 1.0 = 7.9$ mm.

26 Elements of Foundation Design

Estimation of the value of the modulus of elasticity, E

The determination of a realistic value for E is critical in the calculation of ρ_i by the above methods. Such a value is best obtained from in situ tests, such as the pressuremeter or the plate loading test.

It may also be estimated from undrained triaxial tests with the test sample firstly consolidated with a cell pressure of between 0.5 and 0.67 times the in situ effective overburden pressure. The value of E is taken to be the slope of the stress/strain plot, or secant modulus, over the stress range applicable to the problem.

E is also roughly related to the undrained shear strength of clay but the relationship can vary considerably. For normally consolidated clays, E lies somewhere between 500 c_u and 1500 c_u, the lower end of the range applying to clays of high plasticity.

(ii) Consolidation settlement

The consolidation of a clay layer of thickness, H, is given by the expression:

$$\rho_c = \mu m_v H \Delta \sigma_z$$

where μ = the consolidation factor
m_v = coefficient of volume compressibility
$\Delta \sigma_z$ = average increase in vertical pressure within the layer.

The consolidation factor, μ, depends upon both the pore water pressure parameter, A and the shape of the foundation. Skempton and Bjerrum (1957), on the assumption that m_v and A are constant with depth, evolved the equation:

$$\mu = A + (1 - A)\alpha.$$

Values of α are given in table 1.6.

Table 1.6

H/B	Circular footing	Strip footing
0	1.00	1.00
0.25	0.67	0.74
0.5	0.50	0.53
1.0	0.38	0.37
2.0	0.30	0.26
4.0	0.28	0.20
10.0	0.26	0.14
∞	0.25	0

Typical values for μ are:

Soft sensitive clays – possibly greater than 1.0.
Normally consolidated clays – generally less than 1.0.
Average overconsolidated clays – approximately 0.5.
Heavily overconsolidated clays – perhaps as little as 0.25.

In the case of a deep layer of clay only the upper thickness of depth 4B need be analysed. The procedure is to divide the thickness of 4B into a convenient number of horizontal slices and consider the changes in vertical pressure within each one. The total consolidation is taken as the summation of the individual settlements of each slice (see example 1.9).

(b) *Cohesionless soils*

As already discussed, the prediction of the settlement of foundations on sand are based on in situ test results.

(c) *Buoyant, or compensated foundations*

If the predicted settlement of a shallow foundation is too large to be acceptable it is often possible to effect a reduction by placing the foundation at a greater depth.

When the depth of the foundation is such that the weight of excavated soil is equal to the proposed structural load then, at foundation level, the vertical stress on the soil will have the same value both before and after construction. Such a foundation is said to be fully compensated and, in theory, will experience no settlement. It is rarely necessary to design for full compensation as most structures can accept some settlement.

An excavation involves stress relief which can cause swelling in the remaining soil. Swelling effects are negligible in sands but can be of significance in certain types of clay. The extent of swelling is largely dependent upon the availability of moisture and the length of time that the foundation soil is left unloaded. It is therefore necessary, for buoyant foundations in clay, to attempt to keep the excavation dry and to construct the foundation as quickly as possible. If swelling has occurred even a fully compensated foundation will suffer some settlement effects as the soil will recompress when loaded.

There is a physical limit to the depth at which a foundation can be placed. Even if the sides of an excavation can be successfully supported, the weight of material adjacent to the foundation will eventually cause shear failure within the soil.

28 Elements of Foundation Design

Skempton (1951) proposed that the limiting, or critical depth of excavation, z_c, for clays could be obtained from the formula:

$$z_c = N_c \frac{c_u}{\gamma}$$

where c_u = undrained shear strength of soil
 N_c = bearing capacity coefficient (obtained from fig. 1.2)
 γ = unit weight of soil
and that, for a particular depth of excavation, z, the factor of safety against bottom heave was:

$$F = N_c \frac{c_u}{w_s + \gamma z}$$

where w_s = any surcharge load acting on the surface of the soil.

Example 1.8

A building is 50 m square in plan and is to be constructed within a deep bed of soft clay which has an undrained shear strength of 18 kN/m² and a saturated unit weight of 18 kN/m³.

The water table is at the surface of the clay which is overlain by a 1 m thick layer of sand of unit weight 16 kN/m³.

To keep settlement effects within acceptable limits it is necessary to ensure that the net increase in foundation pressure will not exceed 50 kN/m².

Determine the necessary foundation depth if the total applied load will be 300 MN.

$$\text{Gross foundation pressure} = \frac{300 \times 1000}{50 \times 50} = 120 \text{ kN/m}^2$$

$\therefore$ required pressure relief
  due to soil excavation $= 120 - 50 = 70 \text{ kN/m}^2$.

If D is the required depth of the foundation below the top of the clay bed,

$$1 \times 16 + D \times 18 = 70$$

whence $D = 54/18 = 3.0 \text{ m (and } z = 4 \text{ m)}.$

Although this answers the question, it is necessary to check the factor of safety against bottom heave.

$z/B = 4/50 \approx 0$ $\therefore N_c$ (from fig. 1.2) $= 6.2$

$$\therefore F = \frac{6.2 \times 18}{70} = 1.6 \text{ which is satisfactory.}$$

Note. The value of F that can be regarded as satisfactory will depend upon the site conditions, e.g. the standard of workmanship and the time taken to complete construction. For most practical problems F should not be less than 1.2.

Foundations of offshore structures

Introduction

The seventies saw the development of the North Sea oil industry which involved the construction of massive production platforms and the installation of wells up to 10 km deep.

The size of the platforms is difficult to comprehend for anyone not involved in the industry but a dramatic description of a concrete production platform, the Norwegian 'Condeep', was given in the *New Civil Engineer* in 1974. The magazine suggested to its readers that the best way to appreciate the size of 'Condeep' was to think of a foundation as thick as the height of Nelson's column and with a plan size equal to the area of Trafalgar Square. On top of this mammoth slab are placed three concrete columns, each as high as the G.P.O. tower, and on top of these columns is placed a concrete platform, the size of a football pitch.

Design loads

Any structure placed in an offshore location must be designed to withstand the maximum forces it is liable to be subjected to within its lifetime. There is sufficient statistical evidence for the North Sea to be able to predict that the maximum wave liable to occur within a one hundred year period will have the following characteristics:

Height: 24–30 m, depending upon water depth at location.
Period: 14–18 s.
Wavelength: 600 m, giving a wave velocity of some 135 km/h or 85 m.p.h.

If the mean sea level is considered to be the line midway between the trough and the crest of the one hundred year wave then it can be

30 Elements of Foundation Design

accepted that the passage of the crest of the wave will subject the seabed to some pressure increase, Δp, whereas the passage of the trough will subject the seabed to a pressure decrease equal to $-\Delta p$.

At first glance it might be felt that this pressure difference, Δp, will simply be equal to $\frac{1}{2}H\gamma_w$ where H = height of wave and γ_w = unit weight of sea water, but there is a spreading effect within the water which becomes more significant as the depth of water increases. It is now generally assumed that the passage of the 30 m wave will subject the seabed to pressure differences of $\pm \Delta p = 70$ kN/m^2.

These pressure variations are not only large but they occur over a wide area (see fig. 1.10). Bjerrum (1973) stated that, even with dense sand and hard clays, temporary deformations in the order of 25 mm to 50 mm can occur with wave action. Fig. 1.10 is adapted from Bjerrum's paper.

The problem has been analysed using finite element techniques and, for an assumed 160 m depth of sediments, the values of the

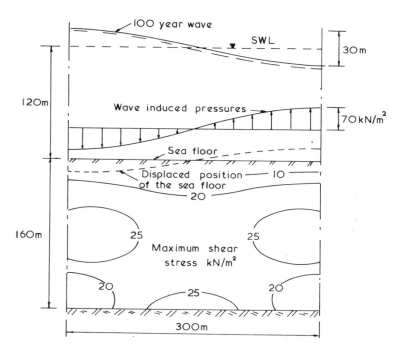

Fig. 1.10 Stresses and displacements in the soil below sea floor when a 30 m wave passes the site (Bjerrum, 1973)

shear stresses to which these sediments will be subjected are shown in fig. 1.10. It is interesting to note that the maximum shear stress increase occurs at some depth below the seabed and illustrates how necessary deep site investigations are.

Eide (1974) indicated typical maximum forces that can be imposed on a large North Sea gravity platform. These forces are shown in fig. 1.11 and are so gigantic that they can hardly be appreciated. For instance the horizontal force, P_H, of 500 MN is equal to the weight of the entire Forth Rail Bridge.

A steel jacket platform presents a much smaller surface area to wind and waves so that force and moment effects are correspondingly reduced. A typical jacket platform will be supported on 9 to 11 piles per leg and the average forces acting on a single pile will be of the order shown in fig. 1.11B.

The basic soil mechanics principles connected with the foundation of an offshore concrete production platform are set out in the remaining pages of this chapter.

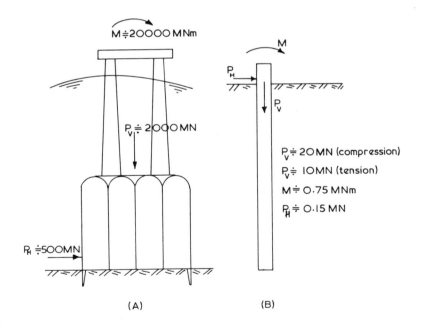

Fig. 1.11 Maximum applied forces on North Sea structures (after Eide, 1974)

32 Elements of Foundation Design

Bearing capacity

For any foundation, the safe bearing capacity is obtained by dividing the ultimate capacity by a suitable factor of safety. For offshore work a factor of 1.5 is usually adopted.

Terzaghi's formula for a shallow (i.e. $z \leq B$) strip footing is:

$$q = \frac{P_v}{B} = c_u N_c + \gamma z N_q + 0.5 \gamma B N_\gamma$$

where P_v = ultimate vertical centrally applied load per unit length.

This is a general formula which assumes that the soil is capable of exhibiting both frictional and cohesive resistance to shear. It should be noted that bearing capacity evaluation is based on the undrained strength characteristics of the supporting soil. Offshore work is involved exclusively with submerged (i.e. saturated) soils. Such soils, whether sand or clay, will act as if they are purely cohesive ($\phi = 0°$) when subjected to undrained shear conditions.

However, due to the high permeability of sand, it is impossible to subject it to undrained shear and therefore, for normal conditions, a foundation on sand is designed on the assumption that the sand is in a drained state (i.e. it has a large ϕ value and $c_u = 0$).

Note. The long term stability of a foundation on clay can always be checked by using the drained strength parameters c' and ϕ' in place of c_u. Normally for offshore work, as for any foundation, this procedure is not necessary, but it may become so for some of the North Sea overconsolidated clays if tests show that they are liable to suffer strength loss with time.

For a surface footing, such as the foundation of a gravity structure in the North Sea, $z = 0$ and N_q is of no account.

For a clay, $\phi = 0°$ and, for offshore work, N_c is generally assumed to have the Fellenius value of 5.14.

For a sand, values for N_γ suggested by Hansen (1970) are shown in fig. 1.12.

To establish a value for N_γ it is first necessary to obtain a value for ϕ. This is obtained from a series of drained triaxial tests carried out on samples of the sand compacted to the in situ relative density.

Note.

$$\text{Relative density, } D_r = \frac{e_{max} - e_{in\ situ}}{e_{max} - e_{min}}.$$

In situ relative density values of an offshore deposit are determined by means of the cone penetrometer, either by being operated from a sunken platform on the seabed, or on the end of a wire line down a borehole.

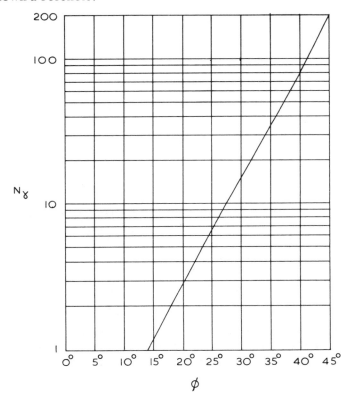

Fig. 1.12 Suggested N_γ values (after Hansen, 1970)

A rough relationship between the cone resistance, relative density and the angle of friction is shown in table 1.7. For a practical problem this table is far too approximate and the penetrometer would have to be calibrated in the laboratory, using remoulded samples of the sand, if an accurate value for D_r is to be obtained for the triaxial tests. Nevertheless the values shown in table 1.7 are useful in that they show the tremendous variation in ϕ, and hence N_γ, with relative density.

34 *Elements of Foundation Design*

Table 1.7

Cone resistance C_r (kN/m²)	Relative density D_r (%)		Angle of friction $\phi°$
0– 1500	Very loose	0–15	30
1500– 4000	Loose	15–35	35
4000–12 000	Medium	35–65	40
12 000–20 000	Dense	65–85	45
> 20 000	Very dense	85–100	> 45

Generally the sand deposits in the North Sea consist of fine sand, densely compacted. At Ekofisk, the cone resistance values varied from 7000 kN/m² to 50 000 kN/m² indicating that the relative density of the sand deposits averaged out at greater than 100%.

Shape factors

Terzaghi's original bearing capacity equation was for a strip footing and this was modified by the use of shape factors so that it could be used for other footings. Terzaghi's original shape factors were:

For sand: $s_\gamma = 0.6$ for a circular foundation
$s_\gamma = 0.8$ for a square foundation.

For clay: $s_c = 1.3$ for circular and square foundations.

The shape factors now generally used for offshore work are those evolved by Hansen and can be used for circular, square and rectangular foundations.

For sands: $s_\gamma = 1 - 0.4 \dfrac{B'}{L'}$

For clays: $s_c = 1 + 0.2 \dfrac{B'}{L'}$

where B' = effective foundation width
L' = effective foundation length.

Effective foundation lengths and widths

The passage of a wave around a gravity platform will subject the structure to a horizontal force, P_H, at some height, h, above the foundation which will therefore be subjected to a moment $M = P_H \times h$. This effect can be allowed for by assuming that the vertical force, P_v, acts at an eccentricity $e \; (= M/P_v)$ from the centre of the foundation (fig. 1.13A). The eccentricity of loading can be allowed for by considering only that part of the foundation that is symmetrical about the assumed point of application of P_v.

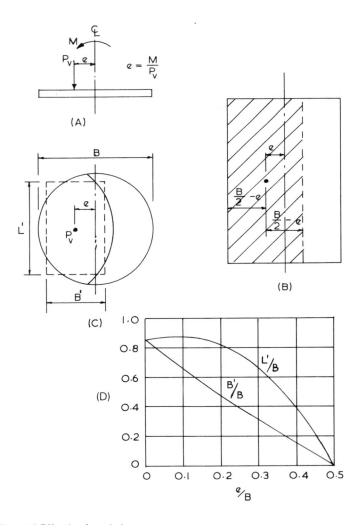

Fig. 1.13 Effective foundation

36 Elements of Foundation Design

For a rectangular footing the problem is relatively straightforward. The equivalent or effective length, L', remains equal to the original length, L and B' is equal to (B − 2e) (fig. 1.13B).

The effective area for a circular foundation is indicated in fig. 1.13C and is treated as a rectangular foundation of dimensions B' and L'.

The relationships between B', L', B and e are shown in fig. 1.13D. (B = diameter).

Inclined load factors

Assuming that P_v acts at an eccentricity, e, from the centre of the foundation does not alter the fact that there is a horizontal force P_H acting on the foundation. The resultant force acting is thus:

$$R = \sqrt{(P_v^2 + P_H^2)} \quad \text{inclined at some angle.}$$

When a foundation is subjected to inclined loading the potential rupture surfaces become unsymmetrical and this leads to a loss of bearing capacity. This loss can be allowed for by a modification to the shape factor by means of a load inclination factor, i_γ. Hansen suggested the following formulae:

For sands: $\quad i_\gamma = \left(1 - 0.7 \dfrac{P_H}{P_v}\right)^5$

and for clays: $\quad i_c = 0.5 - 0.5 \sqrt{\left(1 - \dfrac{P_H}{Ac_u}\right)}$

where $A = B'L'$.

Summary of bearing capacity formulae for offshore conditions

Sand

$$q = \dfrac{P_v}{B'L'} = 0.5\, N_\gamma\, B'\, \gamma'\, s_\gamma\, i_\gamma$$

where γ' = submerged weight of soil (approximately 10 kN/m³)

$$s_\gamma = 1 - 0.4 \dfrac{B'}{L'}$$

$$i_\gamma = \left(1 - 0.7 \dfrac{P_H}{P_v}\right)^5$$

Clay

$$q = \frac{P_v}{B'L'} = 5.14\, c_u\, (1 + s_c - i_c)$$

where $s_c = 0.2\, \dfrac{B'}{L'}$

$$i_c = 0.5 - 0.5\sqrt{\left(1 - \frac{P_H}{Ac_u}\right)}.$$

Sand and clay

In the case of a site consisting of alternating layers of sand and clay it may be possible, at least in the initial design stage, to use the full formula:

$$q = 0.5\, N_\gamma\, B'\gamma'\, s_\gamma\, i_\gamma + 5.14\, c_u\, (1 + s_c - i_c)$$

provided that the separate contributions of the sand and the clay can be assessed.

Whenever it is difficult to allocate an average ϕ or an average c_u value, or when the subsoil conditions consist of variable soil deposits, a rough estimate of bearing capacity can be made by a slip surface analysis.

The soil profile is first drawn out using whatever information is available from the site investigation. A trial rupture surface is then established, with the assumption that the foundation will fail by rotation about one edge. The applied forces tending to cause sliding along the surface are then compared with the cohesive and frictional resistive forces available along it. A factor of safety can therefore be obtained and the procedure is repeated until a minimum factor of safety is found.

The analysis is carried out using an assumed effective foundation to allow for eccentricity and gives a minimum factor of safety, and hence a maximum q value, for a strip footing. Hansen's shape factors must then be applied to any q value obtained by this method.

Skirting

The soil sediments in the North Sea can have large variations in properties over relatively short distances. As a result skirting is an almost essential feature of a gravity platform. It consists of a thin steel or concrete circumferential wall projecting downwards below the underside of the foundation. If there is a soft mud deposit on the surface of the seabed the skirting is intended to penetrate through it. The mud is therefore confined and can be forced to consolidate, through drainage points within the foundation, so that the bearing pressure from the foundation is transmitted down to the firmer soils beneath.

Skirt lengths can vary from about 1.5 m to 3 m, the length being decided upon according to the seabed conditions at the location for the platform. Float-out depths may also dictate the maximum skirting depth.

It should be realised that skirts are not intended to act as bearing piles–the load on the foundation is transmitted through the base slab as bearing pressure. Nevertheless, at penetration, the skirting is subjected to huge axial compressive stresses and must be designed accordingly.

Necessary information

In order to prepare an economical design for the foundation of a large offshore structure it is necessary to know:
(i) The topography of the seabed.
(ii) The nature of the surface sediment.
(iii) The types of soil and their characteristics for the entire depth that will be significantly affected by the structure.
(iv) The forces to which the structure will be subjected.

If the seabed is sloping it may not be possible to use a gravity structure. If, however, there is a surface sediment that can be penetrated by a skirting, and therefore consolidated, a gravity structure may still be possible.

The significantly stressed depth below a gravity structure is roughly $1\frac{1}{2}$ times diameter of foundation ($\approx$ 150 m for North Sea).

For a piled structure the significantly stressed depth is at least the length of the piles.

Example 1.9 Preliminary design for an offshore foundation

Suppose a gravity platform is to be designed with a circular foundation of 100 m diameter. For this design it will be assumed that the maximum forces quoted by Eide (1974) are applicable:

$$\begin{aligned}
\text{Depth of water} &= 150 \text{ m} \\
\text{Design wave: height} &= 30 \text{ m} \\
\text{period} &= 15 \text{ s}
\end{aligned}$$

Horizontal wave force = 500 MN
Submerged weight of structure = 2000 MN
Moment on foundation = 20 000 MNm.

Note. These values are for illustrative purposes only and cannot be considered as being applicable to a particular structure. Actual magnitudes vary with the shape of the structure and hence the resistance it offers to wind and wave forces. Stubbs (1974) illustrated how forces and moments could vary by as much as 100% for different structures.

The main soil problems associated with a gravity structure are:

(a) minimum contact pressure
(b) maximum contact pressure
(c) sliding
(d) overturning (bearing capacity failure)
(e) settlement
(f) cyclic loading effects
(g) design of skirting
(h) scour.

(a) Minimum contact pressure

The 100 year wave will subject the structure to the maximum overturning moment and it is essential that a minimum compressive contact pressure is maintained as the wave passes. If the foundation was to lift at one edge water would be sucked in and out and rapid erosion of the soil would result. If a skirting is employed there may be circumstances when a certain amount of suction could be withstood by the foundation but how much will depend upon the type of soil and the depth of the skirting.

For the structure in this example, ignoring any skirting effects:

$$\text{Area of foundation} = \frac{\pi}{4} \times 100^2 = 7854 \text{ m}^2.$$

A square foundation of the same area, which can be regarded as equivalent, would have a side of:

$$B = \sqrt{7854} = 88.6 \text{ m}$$

and a section nodulus, $Z = \dfrac{B^3}{6} = \dfrac{88.6^3}{6} = 115\,918 \text{ m}^3$.

40 Elements of Foundation Design

$$\text{Bearing pressure} = \frac{P_V}{A} \pm \frac{M}{Z}$$

$$= \frac{2\,000\,000}{7854} \pm \frac{2\,000\,000}{115\,918}$$

$$= 254.6 \pm 172.5 = 427 \text{ or } 82 \text{ kN/m}^2.$$

Minimum pressure is therefore positive and the base will not lift.

(b) Maximum contact pressure

The average bearing pressure (P_V/A) will vary with the type of platform and can be anything from 100 to 600 kN/m². If there are any high spots on the seabed it is possible for a very high local pressure to be applied to the foundation during installation which will be sustained until the high spot flows out in a plastic failure. It has therefore become standard practice, if the topography of the seabed is in doubt (and when is it not?) to design the foundation to withstand upward local pressures of 2000 kN/m².

(c) Sliding

The factor of safety generally used when checking sliding effects is 1.5. Sliding can occur at the interface between the structure and the soil or within the soil at a lower level; skirting more or less ensures the latter case. For illustrative purposes this gravity structure will have no skirting.

Sand

Obviously $\tan \delta = P_H/P_V$

where δ = mobilised angle of friction along potential sliding surface.

If the sliding surface is between the structure and the soil, tests should be carried out to obtain an accurate value for δ. In the absence of tests, δ is often taken as 0.75ϕ, where ϕ is the angle of friction of the sand.

In this example, $\tan \delta = \dfrac{500}{2000} = 0.25$.

Bearing capacity and settlement of foundations 41

Applying a factor of safety of 1.5:

$$\text{required } \tan \delta = 1.5 \times 0.25 = 0.375$$

∴ as sliding surface will be between structure and soil,

$$\text{required } \phi = \frac{\tan^{-1} 0.375}{0.75} = 27.4°.$$

Clay

Shear stress induced by sliding tendency,

$$\tau_H = P_H/\text{area of foundation} = \frac{500\,000}{7854} = 63.7 \text{ kN/m}^2.$$

Generally cohesion between a buried structure and clay is taken as being equal to the undrained shear strength of the clay, c_u. Thus, applying a factor of safety of 1.5:

$$\text{required } c_u = 63.7 \times 1.5 = 96 \text{ kN/m}^2.$$

Note. Where there is a layered soil system beneath the foundation, the strength of the weaker soil is used to determine sliding resistance whereas the strength of the stronger soil is used to estimate skirt penetration resistance.

(d) Overturning (bearing capacity failure)

Sand

Formula is:
$$q = \frac{P_V}{B'L'} = 0.5 \, N_\gamma \, B' \, \gamma' \, s_\gamma \, i_\gamma$$

Effective foundation: $e = M/P_V = \dfrac{20\,000}{2000} = 10 \text{ m}$

$$e/B = 10/100 = 0.1.$$

∴ From fig. 1.13D, $B'/B = 0.68$; $L'/B = 0.88$

$$\therefore L' = 88 \text{ m and } B' = 68 \text{ m}$$

$$s_\gamma = 1.0 - 0.4 \times (68/88) = 0.69$$
$$i_\gamma = (1.0 - 0.7 \times 0.25)^5 = 0.382$$

42 Elements of Foundation Design

$$\therefore \frac{2000000}{68 \times 88} = 0.5 \, N_\gamma \times 68 \times 10 \times 0.69 \times 0.382$$
$$(\text{assuming } \gamma' = 10 \text{ kN/m}^3).$$

Hence $\quad N_\gamma = 3.73$

and, from fig. 1.12, required $\phi = 22°$.
Applying factor of safety of 1.5:

$\tan 22° = 0.404 \quad \therefore \text{required } \tan \phi = 1.5 \times 0.404 = 0.606$

i.e. minimum ϕ for overturning $= 32°$.

Clay

Formula is: $\quad q = \dfrac{P_V}{B'L'} = 5.14 \, c_u \, (1 + s_c - i_c)$

$s_c = 0.2 \, B'/L' = 0.2 \times 68/88 = 0.155$

$i_c = 0.5 - 0.5 \sqrt{\left(1 - \dfrac{500000}{68 \times 88 \times c_u}\right)}$

$\quad = 0.5 - 0.5 \sqrt{\left(1 - \dfrac{83.56}{c_u}\right)}.$

Try $c_u = 100 \text{ kN/m}^2$; then $i_c = 0.297$.
Substituting in formula for q:

$\dfrac{2000000}{68 \times 88} = 5.14 \, c_u \, (1 + 0.155 - 0.297)$

$\therefore c_u = 75.8 \text{ kN/m}^2$ (which is less than trial value).

If a value for c_u less than 83.56 is tried, the quantity beneath the square root becomes negative and i_c must therefore be taken as being equal to 0.5.

Using this value in the formula for q gives $c_u = 99.27 \text{ kN/m}^2$.

Apply factor of safety, required $c_u = 1.5 \times 99.27 = 149 \text{ kN/m}^2$.

Note. Formulae and inclined load factors used in the example are those quoted by the Norwegian authority Det Norske Veritas and may not be accepted by some authorities.

(e) Settlement

Sand
Assume deep layer with $C_r = 20000 \text{ kN/m}^2$.

Consider layer to be 4B m thick. Effective overburden pressure, p'_0, at centre of layer is approximately 2000 kN/m².

Using de Beer and Martens' method with $C_s = 1.5$

$$\frac{C_r}{p'_0} = 1.5 \frac{20\,000}{2000} = 15$$

$$\rho_i = \frac{400}{15} \log_e \frac{2000 + 22.5}{2000} = 0.3 \text{ m}.$$

$\Delta\sigma_z$, 22.5 kN/m², can be evaluated by Steinbrenner's method (see section on consolidation settlement below).

Clay
Immediate settlement.
Soft clay (average E value for North Sea = 10 MN/m²)

$$\rho_i = 0.73 \times \frac{2\,000\,000 \times 100\,(1 - 0.5^2)}{7854 \times 10\,000} \quad \text{(assuming } \mu = 0.5\text{)}$$

$= 1.39$ m.

Hard clay (average E North Sea value = 50 MN/m²)

$$\rho_i = 0.28 \text{ m}.$$

Consolidation settlement.
Soft clay (average values for North Sea; $m_v = 0.2$ m²/MN; consolidation factor $\mu = 1.0$).
Assume layer divided into four equal slices, each 100 m thick.
Size of equivalent square foundation = 88.6 m × 88.6 m (see minimum contact pressure).

Table 1.8

z	m = B'/z	I_σ	$4I_\sigma$	$\Delta\sigma_z = p4I_\sigma$ (kN/m²)
50	0.89	0.164	0.656	167.3
150	0.29	0.036	0.144	36.7
250	0.18	0.015	0.06	15.3
350	0.127	0.01	0.04	10.2

$\rho_c = \mu m_v \Delta\sigma_z H =$

$$\frac{1.0 \times 0.2}{1000} \times 100\,(167.3 + 36.7 + 15.3 + 10.2) = 4.6 \text{ m}.$$

Hard clay (average values for North Sea: $m_v = 0.01$ m^2/MN; consolidation factor $\mu = 0.5$)

$$\rho_c = 0.1 \text{ m}.$$

(f) Cyclic loading effects

As a wave passes around an offshore structure the horizontal force that it exerts acts first in one direction and then in the other. This occurs in a few seconds and subjects the structure to an alternating moment which induces rapid stress reversals in the supporting soil.

It is apparent that the 100 year wave will occur at the height of a storm of several hours duration, after the platform has been subjected to hours of cyclic loading caused by the passage of hundreds of smaller, but significant, waves. It is therefore important that the state of the soil after hours of cyclic loading be taken into account when the passage of the maximum design wave is considered.

The first step is to define a 'design storm'.

It has been found statistically that a North Sea storm so severe that it will only occur once in a hundred years will build up to a maximum over a period of 3 to 9 hours and will then wane over a similar period. To evaluate the effect of cyclic loading, the numbers of waves of various heights that will probably occur over the worst six hours of the storm are obtained. Knowing these wave heights, the various P_H and hence shear stress values can be assessed. Wave heights in a particular time period are defined as the average of the one-third highest waves that occur in that period.

Bjerrum (1973), listed probable wave heights that will occur during the worst six hours of a North Sea storm (table 1.9).

Effects on sand

With sands in the undrained state, cyclic loading gradually causes a build up in the pore water pressure, u. As a sand's strength is dependent upon the value of $\sigma'/\sigma - u$, where σ is the total stress, it is possible for a sand to lose all its stength if u becomes so large that σ' equals zero. At this stage the sand 'liquifies' but failure will occur long before liquefaction; it will occur as soon as the effective stress reduces to a value such that σ' tan ϕ equals the applied shear stress.

Fortunately the sands of the North Sea are generally very dense and liquefaction effects can be ignored. However, with the size of North Sea structures and the permeability of the sand ($k \approx 1 \times 10^{-2}$ mm/s), the rate of pore water dissipation will be much lower than the rate at which it will be generated during a storm and, therefore, undrained conditions, either partial or complete, must be assumed.

Cyclic loading tests are carried out in the laboratory on sand samples in modified shear boxes or triaxial equipment. The pore pressure build up in undrained saturated sand is very slight for a single cycle and it should be remembered that dense North Sea sands, having an extremely small tendency to consolidate, will require many hundreds of cycles before there will be a significant effect on the effective stress. (Loose sands can liquefy after a few cycles.)

A typical result from a cyclic loading test carried out at the Norwegian Geotechnical Institute in Oslo on a sample of densely compacted and saturated fine sand is shown in fig. 1.14. If the rise in pore water pressure per cycle ($\Delta u/N$) is divided by the initial effective vertical stress, σ_c', a parameter, β, is obtained (see fig. 1.14).

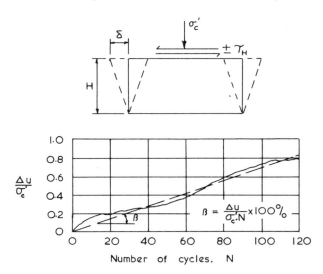

Fig. 1.14 Typical test result of cyclic loading on saturated sand (after Eide, 1974)

It is seen that the pore water pressure increases almost linearly with the number of cycles up to failure (liquefaction).

If a number of cyclic shear tests are carried out at different shear stress levels (τ_H/σ_c') a relationship between stress level and build up in pore water pressure can be obtained. Typical results are shown in fig. 1.15. With this diagram it becomes possible to estimate the total increase in pore water pressure that will occur in sand beneath a gravity structure during the design storm (see example later). It is assumed that these conditions will apply in the soil during the passage of the 100 year wave.

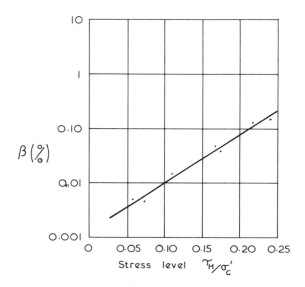

Fig. 1.15 Typical cyclic test results on dense saturated sand

Effect of stress history on sands

The early stress history of a sand layer after the installation of a gravity structure will be important. If the platform is installed in early summer the sand may be subjected to gentle cyclic loading and therefore allowed to densify through consolidation for several months before it is subjected to any storm of note. This effect can be allowed for in the cyclic loading test programme by subjecting the sample to some low stress level ($\tau_H/\sigma_c' = 0.04$) for some 100 cycles and allowing the sample to drain before the true undrained cyclic load test is carried out. This process is known as pre-shearing.

It is interesting to note that very little densification occurs during the pre-shearing of a dense sand and it is thought that the gain in strength may be due more to a structural rearrangement of the grains rather than a reduction in the void ratio.

Effects on clays

If a clay layer occurs on or near the surface of the seabed at a gravity platform location then cyclic loading effects may be important. Generally a layer of soft, sensitive, normally consolidated clay, will behave as a loose sand and suffer large strength reductions under cyclic loading. The presence of such a soil stratum may render a site unsuitable for a gravity structure.

For overconsolidated clays the possible changes in strength with time and with cyclic loading must be investigated. A considerable amount of work still remains to be done in this field although considerable progress has been made so far.

Cyclic loading tests can be conducted in either modified shear or triaxial apparatus.

With the triaxial test the sample is consolidated isotropically with a cell pressure equal to the average effective principal stress value expected in situ. This pressure is maintained during the cyclic loading part of the test, when the sample is tested in an undrained state. The original shear strength of the clay, c_u, is obtained from independent tests on other samples.

The stress levels used in a cyclic load test must be relevant to in situ conditions and therefore must be first determined from a storm/wave analysis. In the triaxial test the stress level is expressed as the ratio between the applied maximum shear stress (i.e. half the value of the applied cyclic deviator stress) and c_u.

After consolidation the deviator stress is increased from zero to one half of its maximum value (see fig. 1.16). This is regarded as the start of the test and the deviator stress is now varied between the maximum and zero values shown in fig. 1.16.

Variation in strain can be obtained by measuring the axial deformations occurring with each cycle. Pore pressure changes can only be obtained after a certain number of cycles if the test is stopped and time allowed for the pore pressures to distribute themselves throughout the sample before a measurement is taken.

However, Theirs and Seed (1969) have indicated that it is possible to predict the reduction in shear strength of a marine clay that occurs during cyclic loading from the strain measurements (fig. 1.17).

It a series of tests are run at different stress levels (i.e. different values of τ_{cyclic}/c_u) it becomes possible to produce a diagram of the form shown in fig. 1.18, which gives the variation of strain with

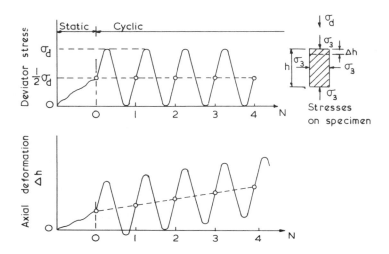

Fig. 1.16 Shear stress and axial deformations in cyclic load triaxial test

stress level and number of load cycles. Shear strain is obtained directly from the shear box or can be taken as the octahedral shear strain ($= 2 \times$ axial strain in an undrained test). With such a diagram and storm/wave information it is possible to predict the amount of strain that will occur within a clay during a storm.

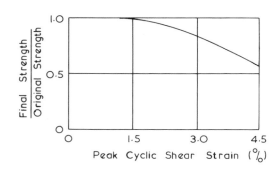

Fig. 1.17 Reduction of clay strength with cyclic loading (after Theirs and Seed, 1969)

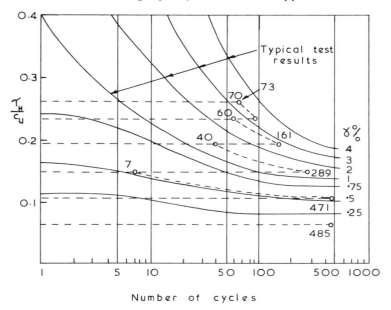

Fig. 1.18 Cyclic tests on clay (after Clausen, 1973)

Returning to our design example:
We have assumed that the design wave will be 30 m high and that it will exert a horizontal force, $P_H = 500$ MN, on to the structure.

Sand

If P_H is assumed to be proportional to the wave that produced it, then, knowing the size of the waves that will occur during a storm and the area of the foundation (7854 m^2), it becomes possible to determine the various values of stress level (τ_H/σ'_c) to which the structure will be subjected.

$$\sigma'_c = \frac{P_v}{7854} = \frac{2\,000\,000}{7854} = 255 \text{ kN/m}^2.$$

Example of stress level determination, using Bjerrum's data for wave height and number (table 1.9):
Wave height = 16 m–20 m, i.e. average height = 18 m

$$P_H = 500 \times \frac{18}{30} = 300 \text{ MN}$$

$$\therefore \tau_H = \frac{300\,000}{7854} = 38.2 \text{ kN/m}^2$$

$$\tau_H/\sigma_c' = \frac{38.2}{255} = 0.150.$$

From the test results of fig. 1.15, the column for β can be completed and the values for Δu obtained.

Table 1.9

Height of waves (m)	Number of waves (N)	Stress level τ_H/σ_c'	Pore pressure per cycle in % of σ_c' (β)	$\beta.N$ (%)
4–8	485	0.050	0.0035	1.7
8–12	471	0.083	0.009	4.24
12–16	282	0.117	0.014	3.95
16–20	121	0.150	0.03	3.63
20–24	32	0.183	0.067	2.14
24–26	3	0.208	0.095	0.29
				13.8

Excess pore water pressure at end of six hour storm:

$$\Delta u = 0.138 \times 255 = 35.2 \text{ kN/m}^2.$$

The numerical value of Δu is of little interest really. What is important is Δu expressed as a percentage of σ_c', in this case it is 13.8%.

In effect therefore, cyclic loading effects will reduce the strength of the sand to 86%. To allow for this, the factor of safety, for both sliding and bearing capacity, should be increased from 1.5 to 1.5/0.86 or 1.75.

This leads to required ϕ values of:

31.5° (for sliding) and 36° (for overturning).

Clay

Stress levels for a cyclic test on clay are quoted as τ_{cyclic}/c_u, not as τ_H/σ_c'. Although τ_{cyclic} is equivalent to τ_H, the stress levels in table 1.9 are no longer applicable as the denominator is now c_u, not σ_c'.

Assuming $c_u = 200 \text{ kN/m}^2$ the stress levels for the same wave data are given in table 1.9a.

Bearing capacity and settlement of foundations 51

Table 1.9a

Height of waves (m)	Number of waves (N)	Stress level τ_{cyclic}/c_u
4–8	485	0.064
8–12	471	0.106
12–16	282	0.149
16–20	121	0.191
20–24	32	0.233
24–26	3	0.265

With this information it is possible to construct the storm path shown in fig. 1.18 from which it is seen that, for this example, the clay will be subjected to a strain of some 3.4%. From fig. 1.17 the reduction in strength that the clay will suffer is found to be about 20% and the undrained shear strength available to resist the effects of the 30 m wave will be reduced to some 160 kN/m².

(g) Design of skirting

It is important that any skirting provided achieves full penetration during installation, an operation that usually takes up to 24 hours.
Penetration resistance is assumed to consist of two components:
(i) Point resistance.
(ii) Friction or adhesion on the walls.
Both components must be assumed to be fully developed and their unit values assumed equal to the maximum values measured during the site investigation programme, the complete opposite procedure used to evaluate sliding and overturning resistance.
Skirtings may be called upon to resist localised high stresses during installation if the seabed is uneven and they must also be designed to withstand shear effects induced when sliding effects are resisted.

(h) Scour

The possibility of scour is of considerable importance in the design of a gravity structure. Because of the inherent increase in current velocity scour can occur around the structure leading to particles of the supporting soil being washed away and perhaps, eventually, to the failure of the structure. The effects of scour are, of course, really of concern only when the foundation soil is sand.

The best protection against scour is a skirting which penetrates the soil to give either partial protection (fig. 1.19A) or complete protection (fig. 1.19B).

In the case of a skirting giving only partial protection, a scour blanket or mattress of plastic material weighed down by gravel or stones can be used to protect the sand close to the structure. This solution is ideal in theory, and worked well at Ekofisk, but it is difficult (and costly) to achieve in deep water.

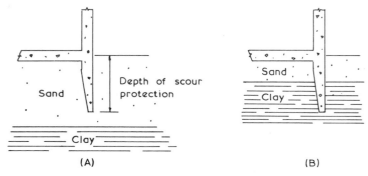

Fig. 1.19 Scour protection

References (chapter 1)

Baguelin, F., Jezequel, J. F. and Le Mehaute, A. (1974) 'Self-boring placement method of soil characteristics measurement.' *Proc. Conf. on Subsurface Exploration.* New Hampshire: Henniker.

Balla, A. (1962) 'Bearing capacity of foundations.' *J. Soil Mech. Fdns Div. Am. Soc. Civ. Engrs* (SM5).

Bjerrum, L. (1973) 'Geotechnical problems involved in foundations of structures in the North Sea.' *Géotechnique.*

Clausen, C. J. F. (1973) 'Stability problems related to offshore gravity structures.' *Paper presented at Offshore Engng Seminar.* Edinburgh, Heriot-Watt University.

De Beer, E. and Martens, A. (1957) 'Method of computation of an upper limit for the influence of the heterogeneity of sand layers in the settlement of bridges.' *Proc. 4th Int. Conf. Soil Mech.*, London.

Eide, O. (1974) 'Marine soil mechanics, applications to North Sea

offshore structures.' *Offshore Technology Conf.* Stavanger.
Gibbs, H. J. and Holtz, W. G. (1957) 'Research on determining the density of sands by spoon penetration testing.' *Proc. 4th Int. Conf. Soil Mech.* London.
Hansen, J. B. (1957) 'Foundations of structures.' General report. *Proc. 4th Int. Conf. Soil Mech.* London.
Hansen, J. B. (1970) 'A revised and extended formula for bearing capacity.' *Geotekniske Institut København Bulletin.*
Janbu, N., Bjerrum, L. and Kjaernsli, B. (1956) 'Vieledning ved Løsning av Fundamenteringsoppgaver.' *N.G.I. Publication No. 16.*
Meigh, A. C. and Nixon, I. K. (1961) 'Comparison of in situ tests for granular soils.' *Proc. 5th Int. Conf. Soil Mech.* Paris.
Menard, L. (1957) 'Mésures in-situ des propriétés physiques des sols.' *Annales des Ponts et Chaussées.*
Meyerhof, G. G. (1951) 'The ultimate bearing capacity of foundations.' *Géotechnique.*
Meyerhof, G. G. (1956) 'Penetration tests and bearing capacity of cohesionless soils.' *J. Soil Mech. Fdns Div. Am. Soc. Civ. Engrs* (SM1).
Meyerhof, G. G. (1965) 'Shallow foundations.' *J. Soil Mech. Fdns Div. Am. Soc. Civ. Engrs* (SM2).
Peck, R. B. and Bazaraa, A. R. S. S. (1969) Discussion in *J. Soil Mech. Fdns Div. Am. Soc. Civ. Engrs* (SM3).
Peck, R. B., Hanson, W. E. and Thornburn, T. H. (1974) *Foundation Engineering.* 2nd edition. Chichester, John Wiley.
Schmertmann, J. H. (1970) 'Static cone to compute static settlement over sand.' *J. Soil Mech. Fdns Div. Am. Soc. Civ. Engrs* (SM3).
Skempton, A. W. (1951) 'The bearing capacity of clays.' *Building Research Congress.*
Skempton, A. W. and Bjerrum, L. (1957) 'A contribution to settlement analysis of foundations on clay.' *Géotechnique.*
Smith, G. N. (1978) *Elements of Soil Mechanics for Civil and Mining Engineers.* 4th edition. London: Granada Publishing.
Steinbrenner, W. (1934) 'Tafeln zur Setzungsberechnung.' *Strasse.*
Stubbs, S. B. (1974) 'Seabed considerations–putting gravity on a firm footing.' *New Civil Engineer Supplement.*
Terzaghi, K. (1943) *Theoretical Soil Mechanics.* Chichester: John Wiley and Sons Inc.
Terzaghi, K. and Peck, R. B. (1948) *Soil Mechanics in Engineering Practice.* Chichester: John Wiley and Sons Inc.

Theirs, G. R. and Seed, H. B. (1969) 'Strength and stress-strain characteristics of clays subjected to seismic loading conditions.' *Spec. tech. Publ. Am. Soc. Test. Mater.* (450), Philadelphia.

Thorburn, S. (1963) 'Tentative correction chart for the standard penetration test in non-cohesive soils.' *Civ. Engng publ. Wks Rev.* **58.**

Wroth, C. P. (1975) 'In situ measurement of initial stresses and reformation characteristics.' *Conf. on In Situ Measurement of Soil Properties.* N. Carolina State University.

2. Piling

Introduction

A pile, just like any other foundation, is intended to transmit a structural load into the ground without risk of shear failure or excessive settlement. There are two basic types of pile, end bearing and friction.

End bearing piles. Generally piles are used to transmit a load through a weak or soft deposit on to a firm stratum at a lower depth which is capable of carrying the load. The most common type of pile is therefore the end bearing pile which derives most of its carrying capacity from the penetration resistance at its foot.

Friction piles, as the name suggests, derive their support mainly from the frictional (or adhesional) resistance generated by the soil around their buried surfaces. An example of a friction pile would be one embedded in a deep layer of clay.

Almost all piles receive support from both bearing and shaft resistance. However, it is generally apparent which component makes the major contribution and it is simpler to refer to the pile as being either end bearing or frictional.

Piles can be made of timber, steel (of various shapes) or concrete (precast or cast in situ).

A brief description of these piles together with notes as to their installation and testing, etc. is given in Smith (1978).

This chapter will deal mainly with concrete piles.

Determination of the bearing capacity of a pile by soil mechanics

The ultimate bearing capacity of a pile, Q_u, is generally represented by the formula:

$$Q_u = Q_b + Q_s - W + \gamma A_b D$$

where Q_b = ultimate bearing resistance available
Q_s = ultimate shaft resistance available
W = weight of pile

γ = average total unit weight of soil
A_b = cross sectional area of pile
D = depth of penetration of pile.

The allowable pile load, Q_a, is found by dividing Q_u by a factor of safety, F, which is intended to guard against excessive settlement.

In the above equation, the term $\gamma A_b D$ indicates the relief in vertical pressure at the base of the pile, as a volume of soil equal to $A_b D$ has to be removed, or displaced, before the pile can be installed in the ground.

If the simplifying assumption is made that the weight of the pile, W, is equal to $\gamma A_b D$ then the equation for Q_u becomes:

$$Q_u = Q_b + Q_s \quad \text{which is its most usual form.}$$

From the point of view of a soils mechanics analysis it is convenient to think of the soil penetrated by the pile as being either totally cohesive or totally granular (i.e. totally cohesionless).

Cohesive soils

The equation $\quad Q_u = Q_b + Q_s$

may be expressed as $\quad Q_u = c_u N_c A_b + c_a A_s$

where c_u = average undrained shear strength of soil at base of pile
N_c = a bearing capacity coefficient
c_a = average adhesion between the pile and the soil
A_s = surface area of embedded length of shaft.

Values of A_b and A_s may be simply computed but some discussion about the other variables in the formula is necessary.

Undrained shear strength, c_u, used to determine Q_b

With relatively homogeneous and intact clay an average value for c_u can be obtained relatively easily from tests on samples of the soil taken from the proposed base level of the pile.

When the strength of a cohesive soil varies rapidly with depth a value for c_u can be obtained by taking the average over the range of from 3B above to 1B below the proposed base level of the pile. (B = pile diameter or width).

The determination of a value for c_u in a fissured clay is particularly difficult as these clays exhibit lower stengths in zones adjacent to the fissures than in the intact material between them. Obviously the larger the size of the sample tested, the greater the chance that the sample contains fissures and, therefore, the greater the chance that the value of c_u obtained will be more realistic.

Bearing capacity factor, N_c

N_c is generally taken as 9.0. In the case of a short pile (D/B < 5.0), however, N_c is usually reduced to the values proposed by Skempton for a circular or square footing (see fig. 1.2).

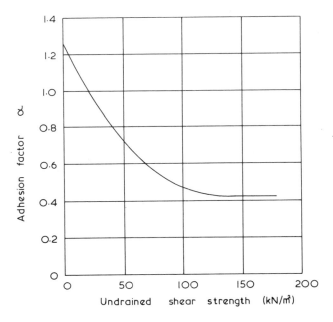

Fig. 2.1 Relationship between the adhesion factor, α, and the undrained shear strength of clays (Tomlinson, 1969)

Average adhesion between pile and soil, c_a

This is generally assumed to be related by the adhesion factor, α, to the undrained soil strength, c_u, such that:

$$c_a = \alpha c_u.$$

The determination of a suitable value for α can be a problem as it not only depends upon the consistency of the soil but also upon whether the pile is to be driven or bored. There are three main methods available:

(a) Tomlinson's method

Driven piles: Tomlinson has prepared a guide as to how α varies with soil strength (fig. 2.1). It can be seen that α can be greater than 1.0 for soft clays but its value falls off rapidly with increasing soil strength.

Bored piles: Skempton (1966) suggested that α can be taken as 0.45.

(b) Meyerhof's method

Driven piles: Meyerhof (1976) suggested that c_u, immediately after driving, is more closely related to the c_u value of the remoulded soil than to that of the undisturbed soil. He proposed that, in the long term, after the dissipation of excess pore water pressures, c_a is controlled by the effective strength parameters of the remoulded soil and may be expressed by:

$$c_a = c' + K_s \bar{p}_0' \tan \phi'$$

where K_s = coefficient of lateral earth pressure
$\bar{p}_0'$ = average effective overburden pressure along pile shaft $\left(= \gamma' \dfrac{D}{2} \right)$.

c' can generally be taken as zero and the expression becomes:

$$c_a = K_s \bar{p}_0' \tan \phi'$$

or $c_a = \beta \bar{p}_0'$ where $\beta = K_s \tan \phi'$ and is known as the skin factor. Fig. 2.2 shows the relationship between β and depth for piles driven into soft and medium clays ($c_u < 100$ kN/m^2).

The above equation for c_a also applies to stiff clays, provided that a realistic K_s can be obtained. With a stiff clay K_s will vary with the overconsolidation ratio which itself varies with depth.

When K_s is unknown for a stiff clay, it can be estimated from the relationship:

$$K_s = (1 - \sin \phi') \sqrt{R_0}$$

where R_0 = overconsolidation ratio (Meyerhof 1976).

Bored piles: Meyerhof maintains that his method for driven piles can also be used for bored piles provided that K_s is taken as 0.8 for stiff clays ($c_u > 100$ kN/m^2) and $(1 - \sin \phi')$ for soft to medium clays ($c_u > 100$ kN/m^2).

(c) Vijayvergiya and Focht's method

Vijayvergiya and Focht (1972) proposed a semi-empirical method that is applicable only to driven piles:

$$c_a = \lambda(\bar{p}'_0 + 2c_u)$$

where λ is a dimensionless factor which varies with the penetration depth of the pile (fig. 2.3).

The approach can be useful for driven piles if there is little information available about the soil's effective strength parameters or overconsolidation ratio.

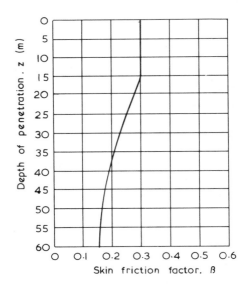

Fig. 2.2 Skin friction factor, β, for piles driven into soft and medium clays (Meyerhof, 1976)

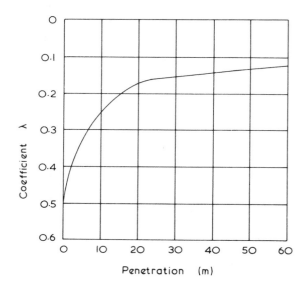

Fig. 2.3 Skin friction coefficient, λ, for piles driven into clay (Vijayvergiya and Focht, 1972)

Note: The following observations should be noted:
(a) Driven piles

(i) Smear effects due to dragdown

When a pile is driven through a soil stratum into an underlying layer some of the material of the upper layer may be dragged down into the lower layer and thus affect the value of the shaft adhesion in that region.

A significant reduction in the value of c_a for the underlying layer can only occur when the upper layer is a soft silt or clay and the lower layer is a stiff to very stiff clay. In such a case it is advisable to reduce the value of c_a, for a depth of 20B below the surface of the stiff clay, to $0.4\,c_u$.

(ii) Annular cracking around top of pile

When a pile is driven through a stiff to very stiff clay deposit that extends from the ground surface downwards a certain amount of cracking and separation of the soil from the pile occurs near its top. To allow for this effect the value of c_a, for a depth from 0 to 20B below the ground surface, should not be assumed greater than $0.4\,c_u$.

(iii) Limiting maximum value for c_a

There appears to be a limit to the maximum amount of adhesion that can be developed between a cohesive soil and a driven pile. It is of the order of 100 kN/m^2 and this value should not be exceeded in bearing capacity calculations, regardless of the proposed depth of penetration of the pile.

(b) Bored piles

(i) Enlarged base

Quite often the bottom of a bored pile is belled out to give an enlarged base and a consequent increase in end bearing resistance. When a pile is enlarged in this manner no shaft adhesion should be allowed for a height of at least two shaft diameters above the top of the bell.

(ii) End bearing in fissured clays

If a bored pile is installed in stiff fissured clay a reduced value of c_u should be used when calculating its end resistance.

A rough rule is:

$$c_u(\text{base}) = c_u(\text{triaxial}) \times 0.75$$

Skempton (1959); Whitaker and Cooke (1966).

Factors of safety for piles in cohesive soils

Driven piles: The safe, or allowable, load for a driven pile in cohesive soil is generally determined from the formula:

$$Q_a = \frac{Q_u}{F}$$

where $F = 2.5$.

Bored piles: With a bored pile a settlement of the order of 0.5%B will fully mobilise the pile's shaft resistance whereas a settlement of about 10%B is required if full end bearing resistance is to be achieved.

Burland *et al.* (1966) suggested that an overall factor of safety be applied to the ultimate pile load (as for a driven pile) but that the value of allowable pile load thus obtained should not be greater than the sum of the fully developed shaft resistance and the ultimate bearing resistance, both divided by partial factors of safety:

62 *Elements of Foundation Design*

i.e. $Q_a = \dfrac{Q_u}{F} \ngtr \dfrac{Q_s}{F_1} + \dfrac{Q_b}{F_2}$.

The values generally used for the factors are

$F = 2.0, F_1 = 1.0$ and $F_2 = 3.0$.

Example 2.1
A single 450 mm diameter concrete pile has been driven 20 m into a soft clay which has an undrained shear strength of 20 kN/m^2 and a unit weight of 18 kN/m^3. Ground water level occurs at the surface of the clay. Determine a value for the allowable pile load.

By way of illustration the ultimate pile load will be determined using each of the three methods described.

$$A_b = \dfrac{\pi \times 0.45^2}{4} = 0.159 \text{ m}^2; \quad A_s = \pi \times 0.45 \times 20 = 28.27 \text{ m}^2$$

(i) Tomlinson's method:

$$Q_u = c_u N_c A_b + c_a A_s$$

$c_u = 20 \text{ kN/m}^2$ ∴ from fig. 2.1, $\alpha = 1.0$ ∴ $c_a = 20 \text{ kN/m}^2$

$Q_u = 20 \times 9 \times 0.159 + 20 \times 28.27 = \underline{504.0 \text{ kN}}$.

(ii) Meyerhof's method:

$$Q_u = c_u N_c A_b + \beta \bar{p}'_0 A_s.$$

Depth of penetration = 20 m ∴ from fig. 2.2, $\beta = 0.28$

$\bar{p}'_0 = (18 - 9.81) \dfrac{20}{2} = 81.9 \text{ kN/m}^2$

$Q_u = 20 \times 9 \times 0.159 + 0.28 \times 81.9 \times 28.27 = \underline{676.9 \text{ kN}}$.

(iii) Vijayergiya and Focht's method:

$$Q_u = c_u N_c A_b + \lambda(\bar{p}'_0 + 2c_u)A_s.$$

Depth of penetration = 20 m ∴ from fig. 2.3, $\lambda = 0.17$

$\bar{p}'_0 = (18 - 9.81) \dfrac{20}{2} = 81.9 \text{ kN/m}^2$

$Q_u = 20 \times 9 \times 0.159 + 0.17 (81.9 + 2 \times 20) \times 28.27 = \underline{614.5 \text{ kN}}$.

Allowable pile load

The three methods have given us three different values for Q_u but, as the methods are all equally acceptable, we are fully entitled to take the maximum of the three values obtained as being Q_u.

Hence $$Q_a = \frac{676.9}{2.5} = \underline{270 \text{ kN}}.$$

Note. When the ground water conditions are unknown, only Tomlinson's formula can be used to predict Q_u.

The reader might like to check that, in example 2.1, if the pile length was made shorter, the method proposed by Vijayergiya and Focht will give a higher value for Q_u than Meyerhof's method.

Example 2.2
A 300 mm diameter concrete pile is driven 15 m into a stiff clay which has an undrained shear strength of 100 kN/m². The unit weight of the clay is 20 kN/m³ and its angle of internal friction, in terms of effective stress, is 30°. If the average overconsolidation ratio, down the length of the pile, can be taken as 5.0, estimate the allowable pile load. The water table can be assumed to be at the surface of the clay.

$$A_b = \frac{\pi \times 0.3^2}{4} = 0.071 \text{ m}^2; \quad A_s = \pi \times 0.3 \times 15 = 14.14 \text{ m}^2.$$

As a value has been given for the overconsolidation ratio Meyerhof's method will be used.

$$K_s = (1 - \sin \phi')\sqrt{R_0} = (1 - 0.5)\sqrt{5} = 1.12$$

$$c_a = K_s \bar{p}_0' \tan \phi' = 1.12\,(20 - 9.81)\,\frac{15}{2} \times 0.577 = 49.39 \text{ kN/m}^2$$

$$\therefore Q_u = 100 \times 9 \times 0.071 + 49.39 \times 14.14 = 762.3 \text{ kN}$$

$$\therefore \text{Allowable pile load}, Q_a = \frac{762.3}{2.5} = \underline{305 \text{ kN}}.$$

Note. The reader might like to check that, in the above example, Tomlinson's method (ignoring R_0) gives $Q_u = 700.2$ kN and Vijayvergiya and Focht's method gives $Q_u = 852.9$ kN. In this example, however, there is no doubt that the correct method to use is the one proposed by Meyerhof.

Example 2.3

A deep bored pile has a shaft diameter of 1 m and an enlarged base of 2.7 m diameter as shown in fig. 2.4

Triaxial tests on representative samples show that the stiff fissured clay, within which the pile is embedded, has an undrained shear strength that varies from 90 kN/m² at its surface to 160 kN/m² at the base of the pile which is 26 m below the surface. Determine the allowable pile load.

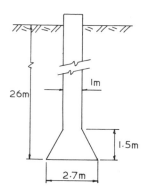

Fig. 2.4

$$A_b = \frac{\pi \times 2.7^2}{4} = 5.73 \text{ m}^2; \quad A_s = \pi \times 1 \times (26 - 1.5 - 2 \times 1)$$
$$= \pi \times 1 \times 22.5 = 70.7 \text{ m}^2.$$

The calculation for A_s assumes no adhesion for twice pile diameter above the bell.

$$c_u \text{ at depth of } 22.5 \text{ m} = 90 + \left(\frac{160 - 90}{26}\right) \times 22.5 = 150.6 \text{ kN/m}^2$$

∴ ultimate shaft resistance = $\alpha \cdot c_u \cdot A_s$

and, using Skempton's value of $\alpha = 0.45$,

ultimate shaft resistance = $0.45 \times \dfrac{(90 + 150.6)}{2} \times 70.7 = 3827$ kN.

Ultimate end resistance =

$$c_u \cdot N_c \cdot A_b = 0.75 \times 160 \times 9 \times 5.73 = 6188 \text{ kN}.$$

A reduced value of 0.75 c_u has been used for the undrained shear strength as the clay is fissured.

Allowable load, $Q_a = \dfrac{3827 + 6188}{2} = 5008$ kN

but this value should not be greater than $\dfrac{3827}{1} + \dfrac{6188}{3} = 5890$ kN

$$\therefore Q_a = \underline{5000 \text{ kN}}.$$

Cohesionless soils

Because of the problem of obtaining undisturbed samples, the design parameters for piles in granular soils are usually obtained from the results of in situ penetration tests.

Peck *et al.* (1953) proposed the following relationship between the uncorrected S.P.T. value, N, and the angle of friction of the granular soil tested:

Uncorrected N value	Relative density of soil	Angle of friction, ϕ, of soil
10	Loose	30°
20 }	Medium dense	33°
30 }		36°
40 }		39°
50 }	Dense	41°
60 }		43°
70 }		44°

The values for the angle of internal friction thus obtained are suitable for most practical problems of piles in granular soils.

End resistance

(i) Driven piles

Meyerhof (1976) proposed that the ultimate end bearing resistance of a pile in a granular soil can be obtained from the expression:

$$Q_b = f_b A_b$$

where f_b is termed the unit penetration resistance of the soil and is calculated from the expression:

$$f_b = p'_0 N_q$$

where N_q = a bearing capacity factor
p'_0 = effective overburden pressure at the pile tip.

The bearing capacity factor, N_q, varies with both the angle of friction of the soil and the penetration depth to width (D/B) ratio (figs. 2.5 and 2.6).

Fig. 2.6 is simply a portion of fig. 2.5 over the range ϕ = 30° to 45°, this being the range within which most practical cases lie.

N_q increases roughly linearly with increasing D/B ratio until it reaches a maximum, and thereafter constant, value at $D_c/2B$. D_c is known as the critical depth and is defined below.

The critical depth, D_c

From the expression $f_b = p'_0 N_q$ it appears that the point resistance of the pile continually increases with depth. Meyerhof pointed out this fallacy and showed that the formula can only apply up to a certain penetration depth, known as the critical depth, D_c.

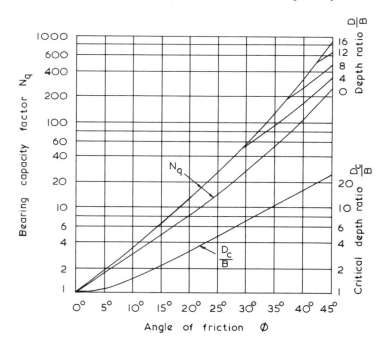

Fig. 2.5 Bearing capacity factor, N_q, and critical depth ratios for driven piles in granular soils (Meyerhof, 1976)

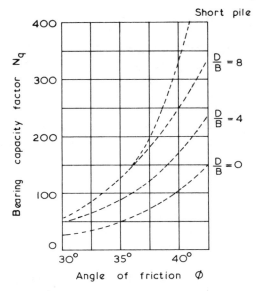

Fig. 2.6 Bearing capacity factor, N_q, for driven piles in granular soils (Meyerhof, 1976)

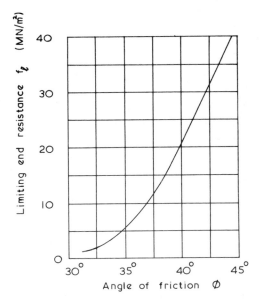

Fig. 2.7 Ultimate end resistance for driven piles in granular soils (Meyerhof, 1976)

68 *Elements of Foundation Design*

Meyerhof's approximate relationship between ϕ and D_c/B is shown in fig. 2.5.

For depths of penetration greater than D_c the bearing resistance of the pile tends towards a constant value and the expression for f_b is rewritten as:

$$f_b = p_0' N_q \leq f_\ell$$

where f_ℓ = limiting value of penetration resistance of the soil (see fig. 2.7). The end resistance for a driven pile achieves a maximum value over a penetration depth range of between 10B and 20B.

End resistance of a pile embedded in an underlying stratum of granular soil

If the pile is driven through a soft deposit, of any material, into a stratum of granular soil lying at a depth greater than the critical depth of the pile, then the end resistance of the granular soil can generally be taken as being at its limiting value f_ℓ (obtained from fig. 2.7).

Note. The pile's critical depth is obtained from fig. 2.5 by assuming that the granular stratum extends up to the surface.

An exception to the above rule occurs when the actual penetration depth into the granular stratum, D_b, is less than 10B. In this case the unit penetration resistance, f_b, should be used in the calculation of the end bearing resistance and f_b should be obtained from the following formula:

$$f_b = f_0 + \frac{(f_\ell - f_0)}{10B} \times D_b$$

where f_0 is an estimated value for any end bearing resistance that might have occurred in the upper layer if the pile had been founded within it (see fig. 2.8).

(ii) Bored piles

The ultimate bearing resistance of a bored pile in a granular soil is generally taken as being $\frac{1}{3}$ to $\frac{1}{2}$ that of the ultimate bearing resistance of a driven pile within the same soil.

Skin friction—bored and driven piles

The average ultimate unit skin friction, f_s, in a homogeneous granular soil can be obtained from the expression:

$$f_s = K_s \bar{p}_0' \tan \delta$$

where $\bar{p}'_0$ = average effective overburden pressure
K_s = coefficient of lateral earth pressure
δ = angle of friction between soil and pile shaft

f_s should not exceed the limiting value f_ℓ.
Tomlinson (1975) quotes approximate values for both K_s and δ which were derived from studies carried out by Broms (1966). These values are given in table 2.1.

Table 2.1 Values for K_s and δ for piles in granular soil

Pile material	δ	K_s Relative density of soil	
		Loose	Dense
Steel	20°	0.5	1.0
Concrete	0.75ϕ	1.0	2.0
Timber	0.67ϕ	1.5	4.0

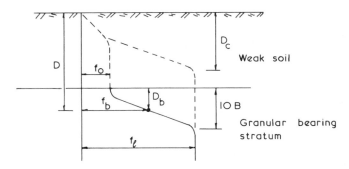

Fig. 2.8 Relationship between ultimate end resistance of driven piles and the penetration depth into a granular soil beneath a weak stratum (Meyerhof, 1976)

Whether the soil into which a pile is installed should be considered as loose or dense depends upon its in situ relative density and the method of pile installation. The larger the volume of soil displaced the higher the value of the resulting shaft friction.

Bored piles should be assumed to be supported by loose soil and driven piles by dense soil. For driven and cast in place piles the soil can be considered as medium dense if the casing is left in or if the concrete is compacted as the casing is withdrawn. If the concrete is

70 Elements of Foundation Design

not compacted than the soil should be assumed to be loose.

Skin friction values calculated by the above formula reach maximum values with penetration depths of between 15 to 20B. It has therefore become standard practice to use this method for depths up to 20B and then to assume that the value thus obtained remains constant with further penetration, with the proviso that f_s should not be greater than 100 kN/m².

Example 2.4

A 10 m long concrete pile, 400 mm diameter, has been driven into a granular soil which has no cohesion and an angle of friction of 33°. The ground water table is at 1 m below the ground surface and the bulk unit weights of the soil above and below the water table are 16 and 20 kN/m³ respectively. Determine the safe bearing capacity of the pile using a factor of safety of 3.

$$B = 0.4 \text{ m}; \phi = 33° \quad \therefore \text{ from fig. 2.5: } D_c/B = 9.0$$

$$\therefore D_c = 0.4 \times 9 = 3.6 \text{ m}$$

i.e. limiting value f_ℓ applies for end bearing.

From fig. 2.7, $f_\ell = 2.8$ MN/m²

$$Q_b = f_\ell A_b = 2.8 \times 1000 \times \frac{\pi \times 0.4^2}{4} = 352 \text{ kN}.$$

Shaft resistance $= f_s A_s$ where $f_s = \bar{p}_0' K_s \tan \delta$
(up to D = 20B)
From table 2.1, $\delta = 0.75 \phi = 25°$
Depth of penetration $= 20B = 20 \times 0.4 = 8$ m
p_0' at 8 m depth $= 16 + (20 - 9.81) 7 = 87.3$ kN/m²
p_0' at 1 m depth $= 16 \times 1.0 = 16$ kN/m².

K_s can vary from K_0 to 4 times K_0 (Meyerhof 1976) and should really be obtained from a pile test. In the absence of data it will be assumed that

$$K_s = 2K_0 \approx 2(1 - \sin \phi') \approx 1.0.$$

Hence, f_s at 1 m depth $= 16 \times 1.0 \times \tan 25° = 7.5$ kN/m²
f_s at 8 m depth $= 87.3 \times 1.0 \times \tan 25° = 40.7$ kN/m²
∴ shaft resistance from upper 1 m of pile $=$

$$\frac{0 + 7.5}{2} \times \pi \times 0.4 \times 1.0 = 4.7 \text{ kN}$$

shaft resistance from remainder of pile =

$$\left(\frac{7.5 + 40.7}{2} \times 7 + 40.7 \times 2\right) \pi \times 0.4 = 314.3 \text{ kN}.$$

Total shaft resistance = 4.7 + 314.3 = 319 kN

$$Q_a = \frac{Q_b + Q_s}{F} = \frac{352 + 319}{3} = 224 \text{ kN}.$$

Example 2.5
A 10 m long concrete pile, 400 mm diameter, is driven through a 7 m thick layer of silty sand having an angle of friction of 25°, into an underlying deep layer of sand whose angle of friction is 35°. The bulk unit weights of the upper and lower layers are 17 and 20 kN/m^3 respectively and the ground water table may be assumed to be at the ground surface. Determine the safe bearing capacity using a factor of safety of 3.

End bearing: Penetration into the lower stratum is 3 m. As this is less than 10B the end bearing will be reduced as shown in fig. 2.8.
Consider first the upper layer where $\phi = 25°$. In the authors' experience the limiting value of end resistance may be taken as varying linearly from 1000 kN/m^2 when $\phi = 30°$ to 400 kN/m^2 when $\phi = 20°$.

$$\therefore f_0 = 700 \text{ kN/m}^2.$$

Consider next the lower layer where $\phi = 35°$. From fig. 2.7 the limiting value of end resistance $f_\ell = 5200$ kN/m^2. This value, however, will not be achieved as $D_b = 3$ m, which is less than the required 10B. In this case:

$$f_b = f_0 + \left(\frac{f_\ell - f_0}{10B}\right) D_b$$

$$= 700 + \left(\frac{5200 - 700}{10 \times 0.4}\right) \times 3 = 4075 \text{ kN/m}^2$$

$$\therefore Q_b = A_b \cdot f_b = \frac{\pi \times 0.4^2}{4} \times 4075 = \underline{512 \text{ kN}}.$$

Skin friction:
(i) Upper silty sand. The limiting skin friction is not reached as the

72 Elements of Foundation Design

thickness of the layer is less than 20B.

$$\therefore f_s = p_0' \cdot K_s \cdot \tan \delta$$

$$\therefore \bar{p}_0' = 7 \left(\frac{17 - 9.81}{2} \right) = 25.17 \text{ kN/m}^2.$$

Taking $K_s = 1$ and $\delta = 0.75 \phi = 18.75°$
$$f_s = 25.17 \times 1 \times \tan 18.75° = 8.54 \text{ kN/m}^2$$
$$\therefore Q_s = A_s \cdot f_s = \pi \times 0.4 \times 7 \times 8.54 = 75 \text{ kN}.$$

(ii) Lower sand layer

$$\bar{p}_0' = 7(17 - 9.81) + \frac{3}{2}(20 - 9.81) = 65.6 \text{ kN/m}^2.$$

$$\therefore Q_s = A_s \cdot f_s = \pi \times 0.4 \times 3 \times 65.6 \times 1 \times \tan 26.25° = 122 \text{ kN}.$$

$\therefore$ Total pile capacity =

$$\frac{Q_b + Q_s}{F} = \frac{512 + (75 + 122)}{3} = \underline{236 \text{ kN}}.$$

Determination of allowable pile loads from in situ tests

In granular soils the values of both f_s and f_b are extremely sensitive to small changes in ϕ. It is possible to obtain reasonable estimates for both these parameters from in situ penetration tests and the following relationships are generally accepted:

Standard penetration test

End bearing

Driven piles in fine to medium sand:

$$f_b = 40N \frac{D}{B} \leq 400 \text{ kN/m}^2.$$

Driven piles in coarse sand or gravel:

$$f_b = 40N \frac{D}{B} \leq 300 \text{ kN/m}^2.$$

Bored piles in any granular soil:

$$f_b = 14N \frac{D}{B} \text{ kN/m}^2$$

where N = uncorrected blow count.

Skin friction

Driven piles (large diameter) $f_s = 2\bar{N}$

Driven piles (average size) $f_s = \bar{N}$

Bored piles $f_s = 0.67\bar{N}$

where $\bar{N}$ is the average of the uncorrected N values along length of pile.

Note. If the borehole was formed by washboring then the average of the N' values should be used (Peck and Bazaraa 1969).

Dutch cone test

The following values have been proposed by Meyerhof (1965) for driven piles:

End bearing

$$f_b = \frac{C_r D_b}{10B} \leq f_\ell \text{ kN/m}^2$$

where C_r is the cone resistance at the pile tip.

Skin friction

$$f_s = \frac{C_r}{200} \text{ kN/m}^2 \text{ for piles in dense sand}$$

$$f_s = \frac{C_r}{400} \text{ kN/m}^2 \text{ for piles in loose sand}$$

$$f_s = \frac{C_r}{150} \text{ kN/m}^2 \text{ for piles in silt}$$

where C_r = average cone resistance over length of pile shaft.

Example 2.6

A 300 mm diameter pile is to be installed by driving it through a 6 m thick layer of normally consolidated soft silt until its tip is embedded at least 1.5 m into an underlying stratum of dense sand. Static cone penetration tests gave the C_r to depth profile shown in fig. 2.9. Determine the allowable load for the pile if the factor of safety is 3.

74 Elements of Foundation Design

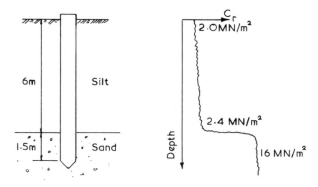

Fig. 2.9

End bearing $= f_b A_b$

where $f_b = \dfrac{C_r D_b}{10B} = \dfrac{16 \times 1000 \times 1.5}{10 \times 0.3} = 8000 \text{ kN/m}^2$

$\therefore Q_b = 8000 \times \dfrac{\pi \times 0.3^2}{4} = 565 \text{ kN}.$

Skin friction (a) In the silt =

$\dfrac{C_r A_s}{150} = \dfrac{2.2 \times 1000}{150} \times \pi \times 0.3 \times 6 = 83 \text{ kN}.$

(b) In the sand $= \dfrac{16 \times 1000}{200} \times \pi \times 0.3 \times 1.5 = 113 \text{ kN}$

$$Q_a = \dfrac{Q_b + Q_s}{F} = \dfrac{565 + 83 + 113}{3} = \underline{254 \text{ kN}.}$$

Negative skin friction

When an end bearing pile has adequate penetration into its bearing stratum (generally $D_b \geq 10B$), it can be considered as incapable of further settlement.

Generally deep deposits of dense sand and/or gravel, stiff clay and rock can be considered as suitable bearing strata.

If, during its installation, a pile passes through a soil deposit which is capable of further settlement then, as consolidation occurs,

the upper regions of this deposit will tend to move downwards.
In the region of the pile, by virtue of the adhesion between it and the soil, the upper surface of the consolidating deposit is prevented from moving downwards and it will develop the shape shown (exaggerated) in fig. 2.10.

As the pile cannot move downwards (apart from any elastic compression) it tends to hold up the adjacent soil, with a consequential increase in its load. The value of this load increase on the pile is equal to the value of unit skin adhesion, or friction, multiplied by the affected area of pile shaft.

This effect of negating some of a pile's allowable load is referred to as negative skin friction.

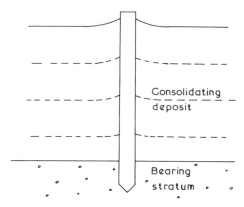

Fig. 2.10 Negative skin friction

The maximum unit value of negative skin friction or adhesion can be determined in exactly the same way as for maximum unit values in friction piles:

For cohesive soils:

Unit negative skin adhesion = $\alpha \bar{c}_u$

where $\bar{c}_u$ = average undrained strength of consolidating deposit
α = adhesion factor, obtained from fig. 2.1 (for driven piles)
= 0.45 (for bored piles).

For cohesionless soils:

Unit negative skin friction = $K_s \bar{p}_0' \tan \delta$ or the value based on C_r.

76 *Elements of Foundation Design*

Important situations where the problem of negative skin friction can occur are:
 (i) Bearing piles driven through recently placed fill material.
 (ii) Bearing piles driven through a soft compressible soil over which a layer of fill material has been placed prior to construction.

Both these cases are discussed later, page 83, but the reader can no doubt think of other instances when negative skin friction effects can take place. For example, if the soft silt deposit in example 2.6 suffered further consolidation due, perhaps, to the application of some highway loading, there could be an increase in the load on the pile and its allowable load would therefore have to be reduced.

The effect can be conservatively estimated if it is assumed that it will occur over the full thickness of the silt when the previously calculated skin friction of 83 kN would become negative.

$$\therefore \text{Allowable pile load} = \frac{Q_b + Q_s}{3} = \frac{565 + 113 - 83}{3} = \underline{198 \text{ kN}}.$$

The effects of negative skin friction become more important when dealing with pile groups and will be discussed again later.

Tension piles

Piles supporting such structures as water towers, electrical transmission pylons and other tall structures, are often required to resist tensile, or pull-out, forces.

When a pile is straight shafted the uplift forces can be assumed to be resisted by shaft friction effects only. The uplift resistance of a pile can be calculated in the same manner as for a pile under compression except that reduced values of the unit skin friction or adhesion should be used. Unless data have been obtained from pull-out tests, unit skin friction and adhesion values should be reduced to one half and even further if the piles are short and are liable to significant loading.

A method for maximising the pull-out resistance of a pile is to bell out its base (fig. 2.11). This has the effect of mobilising a torus of soil above the enlarged base.

Meyerhof and Adams (1968) proposed a solution for the value of the pull-out resistance, P_u.

For a shallow pile:

$$P_u = \pi B \bar{c}_u D + S_f \frac{\pi}{2} B \gamma D^2 K_u \tan \phi + W \qquad \text{(fig. 2.11A)}$$

For a deep pile:

$$P_u = \pi B \bar{c}_u H + S_f \frac{\pi}{2} \gamma B (2D - H) H K_u \tan \phi + W \quad \text{(fig. 2.11B)}$$

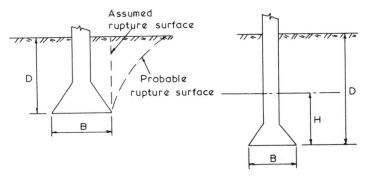

(A) Shallow pile   (B) Deep pile

Fig. 2.11 Uplift resistance of piles

where B = diameter of enlarged base
    $\bar{c}_u$ = average undrained shear strength of soil
    ϕ = angle of friction of soil
    γ = bulk unit weight of soil
    W = weight of soil and pile in cylinder of diameter B and height D
    S_f = a shape factor
    K_u = coefficient of lateral earth pressure
    H = maximum height of rupture surface (see table 2.2).

In the above formulae the first and second terms are the cohesive and frictional components of the uplift resistance and one or other is ignored if the soil is either purely cohesive or purely cohesionless.

According to Meyerhof and Adams, values for K_u and S_f may be assumed to be as follows:

$$K_u = K_p \tan 0.67 \phi$$

(where K_p = coefficient of passive earth pressure)

$$S_f = 1 + \frac{mD}{B}$$

(with a maximum value, for deep piles, of $1 + \frac{mH}{B}$).

Values of m, S_f and H/B are dependent upon ϕ and are given in table 2.2.

Table 2.2

	Values of H/B, m and S_f (Meyerhof and Adams)						
$\phi°$	20	25	30	35	40	45	48
H/B	2.5	3.0	4.0	5.0	7.0	9.0	11
m	0.05	0.10	0.15	0.25	0.35	0.50	0.60
S_f	1.12	1.3	1.6	2.25	3.45	5.5	7.6

Example 2.7

A 500 mm diameter pile has a belled base as shown in fig. 2.12 and is installed in a soil for which $\bar{c}_u = 20 \text{ kN/m}^2$ and $\phi = 20°$.

If the unit weight of the soil is 18 kN/m³ and the unit weight of the pile is 23 kN/m³, determine a value for the allowable pull-out resistance of the pile, assuming a factor of safety of 2.5.

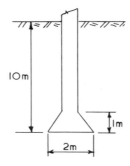

Fig. 2.12

Weight of soil cylinder (if no pile) = $\dfrac{\pi \times 2^2}{4} \times 10 \times 18 = 565 \text{ kN}$.

Weight of concrete (less soil displaced) =

$\dfrac{\pi \times 0.5^2}{4} \times 10 \times (23 - 18) = 10 \text{ kN}$ (ignoring sides of bell)

$\therefore W = 565 + 10 = 575 \text{ kN}$

$\dfrac{D}{B} = \dfrac{10}{2} = 5.0$.

From table 2.2 for $\phi = 20°$ the maximum value of H/B for a shallow footing = 2.5. $\therefore$ Pile must be considered as a deep footing with

$H = 2.5 \times 2 = 5 \text{ m}$.

$K_p = \dfrac{1 + \sin \phi}{1 - \sin \phi} = 2.04 \qquad \therefore K_u = 2.04 \tan 0.67 \times 20° = 0.484$

$P_u = \pi \times 2 \times 20 \times 5 + 1.12 \times \dfrac{\pi}{2} \times 18 \times 2(20 - 5)5$

$\qquad \times 0.484 \times \tan 20° + 575$

$\qquad = 628 + 836 + 575 = 2039 \text{ kN}$

$\therefore$ Allowable pull-out resistance $= \dfrac{2039}{2.5} = 816 \text{ kN}$.

Pile groups

It is rarely that a particular structural need can be satisfied by the use of a single pile. Generally piles are installed in groups as, for instance, in the case of piles supporting a structure. The structure sits on a reinforced concrete slab, called the pile cap, supported on the tops of the piles and distributing the load to them (fig. 2.13C).

The ultimate load carrying capacity of a group of piles cannot always be considered as the summation of the individual carrying capacities of each pile, due to a phenomenon called group action.

When a pile is installed immediately adjacent to another the respective bulbs of vertical pressure can overlap (fig. 2.13B). In the case of a pile group (fig. 2.13C) not only is the underlying soil stressed to a considerably greater depth than with a single pile, but soil outside the perimeter of the group is also stressed. Because of this superposition of vertical stresses at points within the soil it is to be expected that the bearing capacity of a pile group can be less than the summation of the individual pile capacities.

Group action effects are more important with friction piles than with end bearing piles.

Efficiency of a pile group ϵ

The efficiency of a pile group is defined as:

$$\epsilon = \dfrac{\text{ultimate bearing capacity of the group}}{n \times \text{ultimate bearing capacity of a single pile in the group}}$$

where n = number of piles in the group.

80 *Elements of Foundation Design*

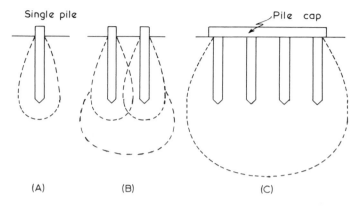

Fig. 2.13 Group action of piles

Ultimate bearing capacity of a pile group in cohesive soil

Except for end bearing piles driven through cohesive soil into an extremely dense bearing stratum, group action effects must be allowed for with a pile group in cohesive soil.

If the piles are placed fairly closely together, failure can occur by the soil shearing around the perimeter of the group. Such a failure is called a block failure.

The ultimate load of a pile group, Q_g, in cohesive soil is therefore the lesser of two possibilities:

(i) $Q_g = \epsilon \times n\, Q_u$ where Q_u = ultimate bearing capacity of single pile
n = number of piles in the group.

(ii) $Q_g = 2D(W + L)\bar{c}_u + c_u N_c WL$

where $\bar{c}_u$ = average undrained shear strength of soil over penetration depth, D
c_u = undrained shear strength of soil at base of piles
N_c = Skempton's value for bearing capacity coefficient (see fig. 1.2)
W = width of pile group
L = length of pile group.

Note. In general, block failure can be avoided if the piles are spaced at centres not less than the perimeter of a pile.

Whitaker (1957) in tests on model pile groups, found that ϵ varied from 0.7 for pile spacings of 2 to 3B, increasing to 1.0 for spacings greater than 8B.

Alternatively, ϵ may be calculated from the Converse-Labarre rule, quoted in several American Building Codes:

$$\epsilon = 1 - \frac{\theta}{90}\left[\frac{(n-1)m + (m-1)n}{mn}\right]$$

where m = number of rows of piles
n = number of piles per row

$$\theta = \tan^{-1}\frac{B}{S} \text{ (in degrees)}$$

B = diameter of piles
S = pile spacing (centre to centre).

Ultimate bearing capacity of a pile group in cohesionless soil

Driving a pile into a cohesionless soil can cause considerable compaction in the adjacent soil. In the case of a group, if the spacings are too close, it can actually become impossible to drive the remaining piles to the required depth, which is why a pile group in cohesionless soil should be driven from the centre outwards.

Due to these compactive effects there can be a considerable increase in soil strength and therefore in the group ultimate load capacity. However the value of this increase is virtually indeterminate and ϵ is generally taken as 1.0 for driven pile groups.

Bored piles in cohesionless soils have no compacting effect and may even create loosening of the soil adjacent to the pile shafts and $\epsilon = 0.7$ is generally adopted.

The above rules assume that the cohesionless soil is of some depth. When there is a weak stratum underlying the pile group a check should be made that the piles cannot punch through the layer of cohesionless soil into the weaker stratum (fig. 2.14).

It has been found that, with driven piles, increase in soil strength due to compaction does not tend to occur with pile spacings equal to or greater than 5B. Generally driven piles are not spaced closer than 2.5B.

For bored piles the minimum spacing is the lesser of either 2B or 750 mm.

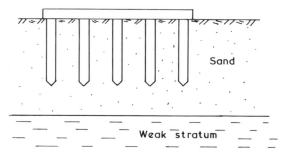

Fig. 2.14 Risk of punching failure

Settlement of pile groups

Due to group action the settlement values of a pile group (both immediate and consolidation) are greater than those for a single pile.

For bearing piles the total foundation load is assumed to act at the base of the piles on an imaginary foundation of the same size as the plan of the pile group (fig. 2.15B).

With friction piles it is virtually impossible to determine the level at which the structural load is effectively transferred to the soil. The level often used in design is at a depth of two-thirds the penetration depth (fig. 2.15A). It is also assumed that there is a spread of the structural load, of one horizontal to four vertical. The settlement of the equivalent foundation is taken as the settlement of the group.

Having established the depth and size of the equivalent foundation, its settlement can be obtained by any standard method, as described in chapter 1, i.e. with the use of soil parameters established from laboratory tests on samples or from the results of in situ tests.

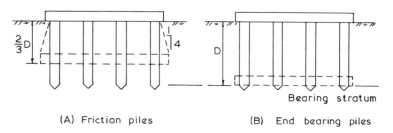

(A) Friction piles   (B) End bearing piles

Fig. 2.15 Equivalent foundations for pile groups

Negative skin friction effects of pile groups

There are two main cases that will be considered:
(1) A group of end bearing piles driven through recently placed frictional fill material that is consolidating under its own weight (fig. 2.16A).
(2) A group of end bearing piles that have been driven through a layer of soft compressible clay over which a layer of fill material had been placed prior to the start of construction (fig. 2.16B).

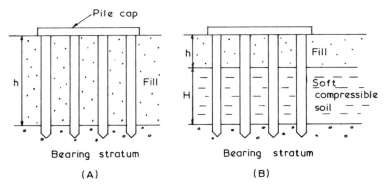

Fig. 2.16 Negative skin friction on a pile group

Case 1
The lower portion of each pile is virtually incapable of downward movement and there are therefore two possible modes of behaviour of the group, depending upon the pile spacing.

(a) Normally spaced piles
Slip will occur between the pile shafts and the consolidating fill and this will induce additional downward forces on to the piles in the form of negative skin friction.

The total induced additional load carried by the group, Q_{group}, because of the negative skin friction effect can be obtained from the expression:

$$Q_{group} = K_s \bar{p}_0' \tan \delta \, (\pi Bh) n \ldots \quad \text{(A)}$$

where n = number of piles and piles are circular of diameter, B.

84 *Elements of Foundation Design*

(b) Closely spaced piles

If the piles are relatively close together the soil within the group will tend to act as a block and little or no slip will occur between the pile and the fill.

If no slip occurs then the entire weight of the fill material will be carried by the pile group and the additional load will equal the weight of the soil.

Weight of soil within the group $\approx LW\gamma h$, where L = length of group; W = width of group; γ = unit weight of fill material.

Around the perimeter of the pile group the surrounding outside fill material will settle which will result in an increase in the induced load carried by the group because of negative skin friction effects.

Increase in downward load due to negative skin friction:

= perimeter of group × negative skin friction
= $2(L + W) K_s \bar{p}'_0 \tan \phi \times h$

where ϕ = angle of internal friction of fill.

Summing these two expressions gives a value for the total induced load on a group of closely spaced piles:

$$Q_{group} = LW\gamma h + 2(L + W) K_s \bar{p}'_0 \tan \phi \times h \ldots \quad (B)$$

The procedure is to evaluate Q_{group} using both expressions (A) and (B) and to take the lesser of the two values as being the additional downward load carried by the pile group.

Case 2

In this instance the soft clay was already consolidating under the weight of the fill material before the piles were driven. This consolidation, possibly augmented by the pile driving, will result in negative skin friction effects.

The approach is similar to that for case 1.

For normally spaced piles the total induced load on the group can be obtained by summing the cohesive and frictional effects of the clay and fill respectively:

$$Q_{group} = \pi Bh \, K_s \bar{p}'_0 \tan \delta \times n + \pi BH \, c_a n \ldots \quad (1)$$

(for circular piles).

For closely spaced piles it must again be assumed that the clay and the fill within the group will act as a block which leads to the alternative expression:

$$Q_{group} = LW\gamma_{fill}h + LW\gamma_{clay}H + 2(L + W)hK_s\bar{p}'_0 \tan \phi_{fill} \\ + 2(L + W)Hc_{u_{clay}} \ldots \quad (2)$$

As before, the lesser of (1) or (2) is taken to be the additional induced downward load acting on the group.

Vertical piles subjected to lateral loading

If a pile is subjected to a horizontal force it is necessary to check that the moments and deflections that will be induced are acceptable.

Various solutions have been proposed for the determination of deflections, moments and shear force values along the length of a pile. It has been found that the manner in which a pile behaves, under the action of a horizontal load, is largely governed by the length of the pile.

With a short pile, failure will be caused by rupture of the surrounding soil, the pile itself remaining undamaged.

With a long pile, failure will be caused by structural damage to the pile which will occur if the value of its yield moment is less than the applied moment.

The deflected forms of short and long piles under the action of a horizontal force applied at ground level are shown in fig. 2.17.

A method of assessing whether a pile is long or short, so that its ultimate load can be determined, is given in example 2.8.

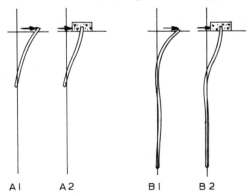

Fig. 2.17 Deflected forms of long and short piles acted upon by a horizontal force. A1: short pile with no head restraint: A2: short pile with pile cap allowing no rotation; B1: long pile with no head restraint; B2: long pile with pile cap allowing no rotation.

86 *Elements of Foundation Design*

Determination of ultimate horizontal load

Broms (1965) has proposed solutions for the determination of ultimate lateral loads on short and long piles in both granular and cohesive soils.

Figs. 2.18A and B represent his solutions for short piles installed in cohesive and granular soils respectively. Note that, in fig. 2.18B, K_p is the coefficient of passive earth pressure and γ is the bulk unit weight of the soil.

With increasing pile length the lateral resistance becomes more closely allied to the pile's moment of resistance (i.e. yield moment).

Figs. 2.19A and B represent Brom's solutions for long piles installed in cohesive and granular soils respectively.

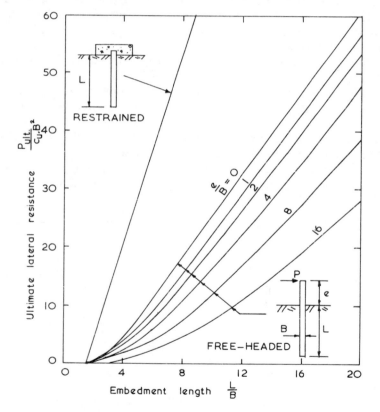

Fig. 2.18A Ultimate lateral load, P_{ult}, for short piles in cohesive soil (D/B < 20) (Broms, 1965)

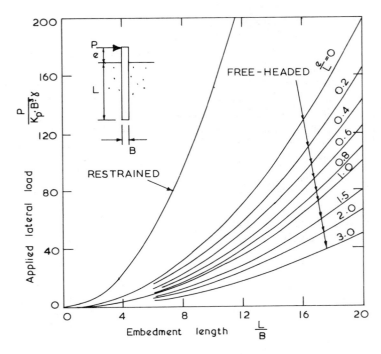

Fig. 2.18B Ultimate lateral load, P_{ult}, for short piles in granular soil ($D/B < 20$) (Broms, 1965)

Determination of displacement at top of pile

The displacement at the top of a pile can be obtained from fig. 2.20A (for a cohesive soil) and fig. 2.20B (for a granular soil). In fig. 2.20A, the deflection has been plotted as a function of the dimensionless length L in which

$$\beta = \left(\frac{k_h B}{E_p I_p}\right)^{\frac{1}{4}}$$

where E_p = modulus of elasticity of pile material
I_p = moment of inertia of pile in plane of bending
B = pile diameter or width
k_h = coefficient of horizontal subgrade reaction (see chapter 3).

88 Elements of Foundation Design

Much of the accuracy of the calculated value of displacement depends upon the accuracy of the assumed value for k_h.

Ideally, horizontal plate load tests or pressuremeter tests should be carried out and, for major works, a full scale pile test involving the installation and jacking apart of two piles of the type proposed for the foundation is necessary if accurate data are to be obtained.

In the absence of such information it is necessary to estimate a value. Broms, using Terzaghi's earlier work (see chapter 3) considered that k_h for a cohesive soil can be considered as constant with depth but, for a granular soil, k_h increases with depth in accordance with the formula:

$$k_h = \frac{n_h z}{B}$$

where z = depth of point considered
and n_h = constant of horizontal subgrade reaction for piles in granular soil.

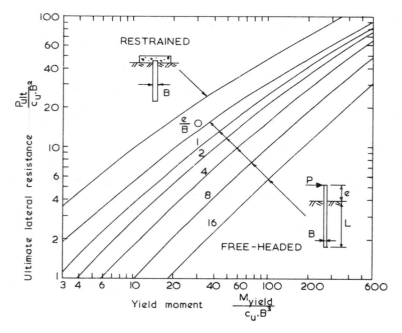

Fig. 2.19A Ultimate lateral load, P_{ult}, for long piles in cohesive soil (D/B > 20) (Broms, 1965)

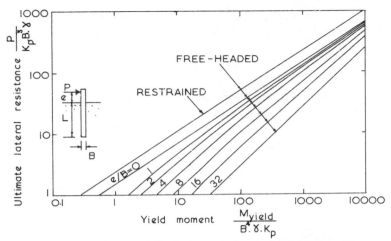

Fig. 2.19B Ultimate lateral load, P_{ult}, for long piles in granular soil (D/B > 20) (Broms, 1965)

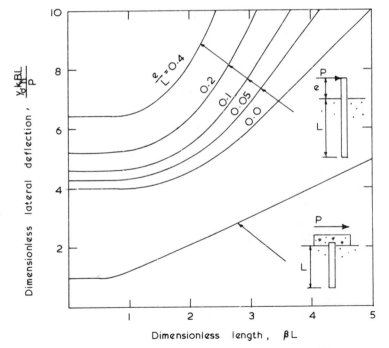

Fig. 2.20A Lateral deflection of pile at ground surface (cohesive soil) (Broms, 1965)

In fig. 2.20B (for granular soils) the relative stiffness of the pile and the soil are contained in the parameter η where:

$$\eta = \left(\frac{n_h}{E_p I_p}\right)^{\frac{1}{5}}$$

Typical values for n_h (for granular soils) and k_h (for cohesive soils) are given in table 2.3.

Table 2.3 Typical values for n_h and k_h

	n_h (MN/m³) for granular soils		
	Loose	Medium	Dense
Dry or moist	2.2	6.6	17.5
Submerged	1.25	4.4	10.5

	k_h (MN/m³) for cohesive soils
Soft	8.0
Medium	16.0
Stiff	32.0

An alternative estimation of k_h (for cohesive soils) as proposed by Vesic (1961) is:

$$k_h = \frac{0.65}{B} \sqrt[12]{\left(\frac{E_s B^4}{E_p I_p}\right)} \times \frac{E_s}{1 - \mu^2}$$

where μ = Poisson's ratio of the soil
E_s = Elastic modulus of the soil
B = Width or diameter of pile
E_p = Elastic modulus of pile material
I_p = Moment of inertia of pile.

An approximation for E_s, which can be used in the above formula, can be obtained from the coefficient of compressibility of the soil, m_v, using the following equation proposed by Francis (1964):

$$E_s \approx \frac{3(1 - 2\mu)}{m_v}.$$

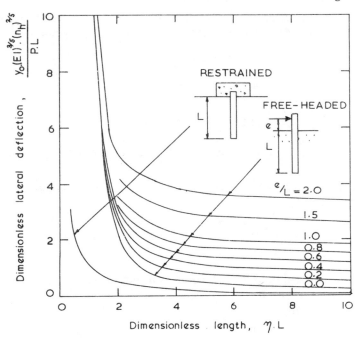

Fig. 2.20B Lateral deflection of pile at ground surface (granular soil) (Broms, 1965)

Example 2.8

A reinforced concrete pile 6 m long by 500 mm diameter is driven into a cohesive soil which has an average undrained shear strength of 30 kN/m². The pile is subjected to a horizontal load of 50 kN acting 1 m above the surface of the soil. Check the suitability of the pile to withstand this load and determine the horizontal deflection of the pile at ground level. Take $E_p = 21 \times 10^6$ kN/m²; $k_h = 7.1$ MN/m³.

Note. At the start of the calculations it is not known whether the pile is long or short. The design procedure adopted here is to split the calculations into two parts:
 (i) Knowing the proposed length of the pile and using fig. 2.18, check whether or not there is adequate lateral resistance if the pile acts as a short pile. If failure is indicated then the pile must be changed in dimension until it is satisfactory.
 (ii) Once it has been established that the pile is satisfactory under (i) then its behaviour as a long pile must be checked.

By means of fig. 2.19 the pile's moment of resistance, or yield moment, can be compared with the value of the applied moment.

$$e/B = 1/0.5 = 2.0 \qquad L/B = 6/0.5 = 12$$

$\therefore$ From fig. 2.18A: $\dfrac{P_{ult}}{c_u B^2} = 27$

$\therefore P_{ult} = 27 \times 30 \times 0.5^2 = 202.5$ kN.

This gives a factor of safety $= \dfrac{202.5}{50} = 4.05$ which is satisfactory.

From fig. 2.19A the required ultimate yield moment can be obtained:

$$\dfrac{M_{ult}}{c_u B^3} = 120$$

$\therefore M_{ult}$ required $= 120 \times 30 \times 0.5^3 = 450$ kNm.

This value should now be compared with the calculated moment of resistance of the pile in accordance with reinforced concrete design principles.

To obtain the value of the deflection at the ground surface it is necessary to evaluate β.

Now $\quad I_p = \dfrac{\pi r^4}{4} = \dfrac{\pi \times 0.25^4}{4} = 3.068 \times 10^{-3}$ m^4

$\therefore \beta = \left(\dfrac{7.1 \times 1000 \times 0.5}{21 \times 10^6 \times 3.068 \times 10^{-3}} \right)^{\frac{1}{4}} = 0.485.$

Now $\quad e/L = 0.167$ and $\beta L = 0.485 \times 6 = 2.91$.

From fig. 2.20A, $\dfrac{y_0 k_h BL}{P} = 8$

$\therefore y_0 = \dfrac{8 \times 50}{7100 \times 0.5 \times 6} = 1.88 \times 10^{-2}$ m.

Deflection of pile at ground surface = 18.8 mm.

Numerical methods

More rigorous and flexible solutions for laterally loaded pile problems can be obtained by the use of a computer which can be programmed to deal with varying pile stiffness and unusual variations in soil resistance. Bowles (1977) lists several computer programs of this type.

A straightforward numerical method, employing the principle of subgrade reaction modulus in which deflections and moments in a laterally loaded pile may be obtained, is described in chapter 3.

Eccentric and inclined loads on pile groups

There are occasions when a horizontal force acting on a pile group is accompanied by a vertical load, due to the weight of the pile cap or some supported structure. In this case we have the problem of a pile cap acted upon by an eccentric inclined load which will distribute itself into the piles.

If there is no horizontal load and if the vertical load is concentric with the centroid of the pile group, the load in each pile is simply taken to be equal to the total load divided by the number of piles. piles.

i.e. $$Q_p = \frac{Q_v}{n}$$

where Q_v = total vertical load
Q_p = vertical load per pile
n = number of piles.

Now consider the case of a vertical load which is eccentric (fig. 2.21).

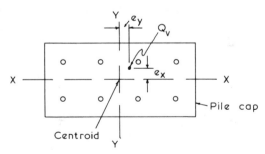

Fig. 2.21 Pile group with eccentric vertical load

Consider a pile in the group at distances x and y from the centroid of the group. Remembering, from the theory of beam bending, that tensile and compressive stresses at a distance, y, from the neutral axis are given by:

$$f_b = \pm \frac{My}{I}$$

where M = applied moment and I = moment of inertia of beam section, the total vertical load induced in the pile can be expressed by:

$$Q_p = \frac{Q_v}{n} \pm \frac{A_b y M_x}{I_x} \pm \frac{A_b x M_y}{I_y}$$

where A_b = cross sectional area of pile
I_x = moment of inertia of pile group about XX axis
= $I_0 + A_b \Sigma y^2 = A_b \Sigma y^2$ since I_0, the moment of intertia of the pile section is negligible.

Similarly $I_y = A_D \Sigma x^2$
$M_x = Q_v e_y$ and $M_y = Q_v e_x$.

Hence $Q_p = Q_v \left(\dfrac{1}{n} + \dfrac{x e_x}{\Sigma x^2} + \dfrac{y e_y}{\Sigma y^2} \right)$

(where e_x and e_y include the algebraic sign).

When a horizontal load acts in conjunction with a vertical load the determination of Q_p for a particular pile is carried out as just described but it is much more difficult to decide what proportion of the horizontal load should be allotted to a pile.

Even with a rigid pile cap, the piles nearest to the point of application of the horizontal load will tend to be displaced more than piles further away (due to elastic compression within the pile cap).

General practice is to assume that the horizontal load is divided equally amongst the piles and to use this value of H/n along with the maximum value of Q_p in order to check the suitability of the piles.

Example 2.9
The plan of a group of nine piles is shown in fig. 2.22. A load of 3250 kN, Q, inclined at 10° to the vertical, acts in a direction parallel to the X–X axis and its point of application has eccentricities, e_x = 0.5 m; e_y = 0.7 m.

Determine the values of vertical load and horizontal load that should be used to check the suitability of the piles.

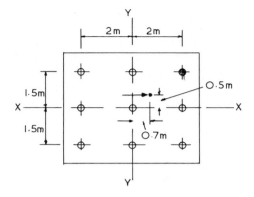

Fig. 2.22

Horizontal component of inclined load = 3250 × sin 10° = 564 kN.
Vertical component of inclined load = 3250 × cos 10° = 3200 kN.

By inspection it is seen that the most heavily loaded pile will be the one shown shaded in the figure. For this pile $x = 2$ m and $y = 1.5$ m.

$$\Sigma x^2 = 6 \times 2^2 = 24; \quad \Sigma y^2 = 6 \times 1.5^2 = 13.5.$$

$$Q_p = 3200 \left(\frac{1}{9} + \frac{2 \times 0.5}{24} + \frac{1.5 \times 0.7}{13.5} \right) = 737 \text{ kN}.$$

Horizontal load per pile $= \dfrac{564}{9} = 63$ kN.

Hence piles should be checked that they can each support a vertical load of 737 kN and a horizontal load, acting at level of underside of pile cap, of 63 kN.

Raking (or battered) piles

When the value of a horizontal force acting on a group of vertical piles becomes too large to be accommodated a common solution is to provide some sloping (or battered) piles along with the vertical ones. For ease of analysis it is generally assumed that the vertical piles resist the vertical load and the inclined piles resist the horizontal load which is obviously extremely conservative.

Piles are not generally inserted into the ground at batters in excess of 4 vertical to 1 horizontal as to do so requires specialist equipment with consequent increased costs.

96 Elements of Foundation Design

Terzaghi (1943) suggests that a graphical method, described by Lohmeyer (1930) and known as Culmann's method, can be used in the analysis of a mixed (i.e. vertical and inclined) pile group.

Often pile groups designed to resist horizontal forces that can act in either direction, for example piles supporting quay walls, are made up of one set of vertical and two sets of inclined piles which are battered in opposite directions (fig. 2.23A).

If each set of piles is replaced by an imaginary pile, located at the centre line of the set, a simple model consisting of three imaginary piles is created (fig. 2.23B).

Assuming that these imaginary piles are acted upon by axial forces only and that they are pin jointed at both ends it becomes possible to determine the positions of the lines of action of the forces that act upon them (fig. 2.23C).

Let Q_A, Q_B and Q_C be the forces per unit length of foundation in the imaginary piles A, B and C.

Let the resultant of Q_B and Q_C be R'.

Then R' must pass through a, the point of intersection of Q_A and the applied load R.

Hence Q_A, Q_B and Q_C can be obtained from a force polygon (fig. 2.23D).

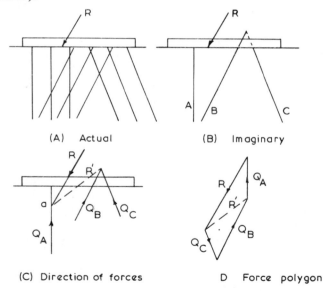

Fig. 2.23 Culmann's method for raked piles

∴ The axial force in each pile in set A, $Q_{pA} = \dfrac{Q_A}{n_A}$

where n_A = number of piles in set A.

Similarly: $Q_{pB} = \dfrac{Q_B}{n_B}$; $Q_{pC} = \dfrac{Q_C}{n_C}$.

References (chapter 2)

Bowles, J. E. (1977) *Foundation Analysis and Design*. 2nd edition. Maidenhead: McGraw-Hill.

Broms, B. B. (1965) 'Design of laterally loaded piles.' *J. Soil Mech. Fdns Div. Am. Soc. Civ. Engrs* (SM3).

Broms, B. B. (1966) 'Methods of calculating the ultimate bearing capacity of a pile.' *Sols-Soils* **18–19**.

Burland, J. B., Butler, F. G. and Dunican, P. (1966) 'The Behaviour and design of large diameter bored piles in stiff clay.' *Conf. on Large Bored Piles*. London: Instn Civ. Engrs.

Francis, A. J. (1964) 'Analysis of pile groups with flexural resistance.' *J. Soil Mech. Fdns Div. Am. Soc. Civ. Engrs.* (SM3).

Lohmeyer, E. (1930) 'Die Berechnungverankerter Bohlwerke.' *Bautechnik* **8**.

Meyerhof, G. G. (1965) 'Penetration tests and bearing capacity of cohesionless soils.' *J. Soil Mech. Fdns Div. Am. Soc. Civ. Engrs* (SM1).

Meyerhof, G. G. (1976) 'Bearing capacity and settlement of pile foundations.' *J. Geotech. Engng Div. Am. Soc. Civ. Engrs* (GT3).

Meyerhof, G. G. and Adams, J. I. (1968) 'The ultimate uplift capacity of foundations.' *Canadian Geotech. J.* **5**(4).

Peck, R. B. and Bazaraa, A. S. (1969) 'Settlement of spread footings on sand.' *J. Soil Mech. Fdns Div. Am. Soc. Civ. Engrs* (SM3) (Discussion).

Peck, R. B., Hanson, W. E. and Thornburn, T. H. (1974) *Foundation Engineering*. 2nd edition. Chichester: John Wiley and Sons Inc.

Skempton, A. W. (1951) 'The bearing capacity of clays.' *Building Research Congress*.

Skempton, A. W. (1959) 'Cast in situ bored piles in London clay.' *Géotechnique* **7**.

Skempton, A. W. (1966) 'Summing up.' *Conf. on Large Bored*

Piles. London: Instn Civ. Engrs.
Smith, G. N. (1978) *Elements of Soil Mechanics for Civil and Mining Engineers*. 4th edition. London: Granada Publishing.
Terzaghi, K. (1943) *Theoretical Soil Mechanics*. Chichester: John Wiley and Sons Inc.
Tomlinson, M. J. (1975) *Foundation Design and Construction*. 3rd edition. London: Pitman.
Vesic, A. S. (1961) 'Beams on elastic subgrade and the Winkler hypothesis.' *Proc. 5th Int. Conf. Soil Mech.* Paris.
Vijayvergiya, V. N. and Focht, J. A. (1972) 'A new way to predict capacity of piles.' *4th Offshore Technology Conf.* Houston.
Whitaker, T. (Dec. 1957) 'Experiments with model piles in groups.' *Géotechnique* 7.
Whitaker, T. and Cooke, R. W. (1966) 'An investigation of the shaft and base resistance of large bored piles in London clay.' *Conf. on Large Bored Piles*. London: Instn Civ. Engrs.

3. Finite Difference Numerical Techniques for Foundations

Introduction

A few years ago many large and complicated structures were mainly designed by rule of thumb methods. Although the designers knew the theory governing the performance of their structures, there simply was not enough time available to solve the various differential equations.

The abilities of modern day computers mean that the same type of civil engineering problem can now be analysed by sophisticated mathematical methods. In the field of geotechnics most of these problems have been written as computer programs, see for example Bowles (1974).

This development should lead to safer and more economical structures but there is a danger that the consultant responsible for the design must lean heavily on the computer expert (unless he has the time to become one himself). This may lead to a loss of some of the control he had in the past.

With the advent of programmable calculators it has become possible, in a number of cases, for the designer at least to check out some of the computer results and to analyse some of the work himself.

The aim of this chapter and the appendices is to present the reader with enough material to analyse foundation problems by numerical methods, with the aid of a calculator. All the worked examples have been solved using a T.I. programmable 59 calculator.

Note. In order to keep this chapter to a reasonable length it has been assumed that the reader has a working knowledge of matrix algebra. For those readers who have not, or who feel in need of revision, the relevant material used in this chapter is set out in Appendix I. Hopefully it should not prove too difficult to work through.

Foundations

The philosophy of foundation design

The function of a foundation is to spread its loading, whether it is from concentrated loads, P, or distributed loads, p, into the supporting soil or rock beneath it. This spreading action creates upward contact pressures, q, between the soil and the underside of the foundation (fig. 3.1).

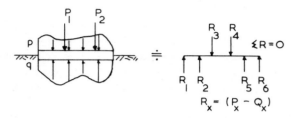

Fig. 3.1

For any section along the foundation the upward pressure values, q, can be considered as equivalent to an upward reactive force, Q. Whilst the upward reactive forces and the downward applied loading are in overall equilibrium, it is their imbalance, R, at sections along the foundation that causes the bending moments and shear forces within it. Knowing the values of shear force and bending moment at critical sections, it becomes possible to design the foundation (i.e. to decide upon its dimensions and amount of reinforcement required).

The design of a foundation therefore resolves into obtaining a reasonable estimation of the contact pressure distribution, q, between it and the supporting soil.

The form of contract pressure distribution beneath a foundation depends upon various factors, the principal ones being:

The form, or shape, of the foundation

The degree of rigidity of the foundation

The type of soil

The form of the applied loading.

Considering each of these factors in turn:

Form of the foundation

Whenever possible, it is an advantage to use foundations of regular shape. For most practical cases a foundation will be either circular, square, rectangular or trapezoidal.

Rigidity of foundation

This aspect of foundation design is fully decribed in Appendix II. At this point all that need be said is that a rigid foundation is unable to bend whereas a flexible foundation offers no resistance to bending and its deflected shape follows any depressions in the surface of the supporting soil. Most foundations lie somewhere between the two extremes.

Type of soil

Soils are basically of two types:

Cohesive–saturated silts and clays which tend to act elastically when loaded, i.e. they suffer a change in shape but not in volume. Hence, if time and consolidation effects are ignored, a foundation placed on such soil can be analysed by elastic theory.

Cohesionless–gravels and sands, either dry or submerged. These soils do not act elastically when supporting a surface footing.

Form of loading

The form of loading to which a foundation will be subjected is dependent upon the size and function of the supported structure. The assessment of the magnitude of these forces is an important part of the design, particularly deciding upon what proportion of imposed loading should be assumed to act on the foundations. With cohesionless soil little harm, other than loss in economy, will occur if there is an overestimation of a particular column load, but with cohesive soil, such an overestimation could lead to differential settlement problems.

Standard solutions

Various solutions for the contact pressure distribution beneath regularly shaped foundations are available. These can be useful in routine design work but they have various disadvantages:

(i) They deal only with simple loading cases

(ii) The soil must be taken as either perfectly cohesive or perfectly cohesionless.

(iii) The foundation must be assumed to be either perfectly flexible or perfectly rigid.

(iv) Solutions are applicable only to:
Square or circular foundations

The transverse direction of strip foundations of regular cross section.

Flexible foundations

These have no resistance to bending and, no matter what the soil type, the contact pressure distribution will always be uniform for concentric loading or uniformly varying for eccentric loading (fig. 3.2A).

The deflected form of a flexible footing, after being loaded, will vary with the type of soil on which it sits. With a surface footing on sand there is no overburden pressure at its edges to give the sand any shear strength whereas the sand below the centre of the foundation rapidly gains strength as soon as loading is applied. The result is that settlement around the edges of the foundation is much larger than that at the centre and the deflected form of the foundation tends to be convex upwards (fig. 3.2B). With a surface footing on clay, with a strength independent of any overburden pressure, the settlement at the centre can be up to $1\frac{1}{2}$ times that at the edges and the deflected form of the foundation tends to be concave upwards (fig. 3.2B).

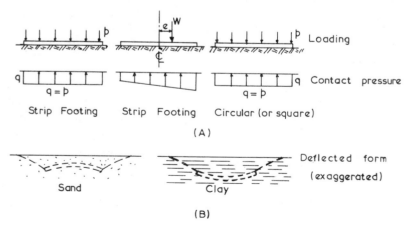

Fig. 3.2

Rigid foundations

If a foundation is perfectly rigid the rigidity will impose an equal settlement beneath the foundation for uniform loading or a lineal variation of settlement for non-uniform loading (fig. 3.3C).

This situation means that, unlike a flexible footing, the contact pressure beneath a rigid foundation depends upon the soil type.

Cohesive soil

A rigid circular foundation on cohesive soil, loaded with a uniformly distributed pressure, p, will experience a contact pressure beneath its centre of about 0.5p. In theory the value of the contact pressure increases parabolically to infinity at the foundation edges. However, as the maximum value of q cannot exceed the shear strength of the clay, a certain amount of local yielding takes place so that the probable contact pressure distribution will be more as shown in fig. 3.3A.

Cohesionless soil

As with the flexible footing, the sand immediately beneath the centre of a rigid surface foundation can attain a considerable strength whilst the soil at the edges remains with virtually no strength. The maximum resistance to compression is therefore at the centre of the foundation and this resistance reduces to zero at the edges. Due to the foundation rigidity a uniform settlement is imposed on the soil so that large contact pressure values are generated beneath the centre and these reduce, approximately parabolically, to zero at the edges (fig. 3.3B).

For a given value of applied pressure the contact pressure distribution varies with the size of the footing tending to become less peaked at the centre as the width of the footing increases (fig. 3.3B). In the case of a deep rigid footing the soil at the edges has some strength so that the contact pressure distribution tends to be as shown in fig. 3.3B.

Routine design calculations

For routine work, involving the design of square or circular pad foundations most designers assume that the contact pressure distribution is either uniform or varies uniformly.

All the solutions just described are for limiting cases, the

104 Elements of Foundation Design

foundation is either flexible or rigid, the soil is either cohesive or cohesionless. All foundations have some degree of flexibility and most soils are a mixture so that, for most practical cases, the actual contact pressure distribution will be a merger of the two limiting forms, tending towards uniformity.

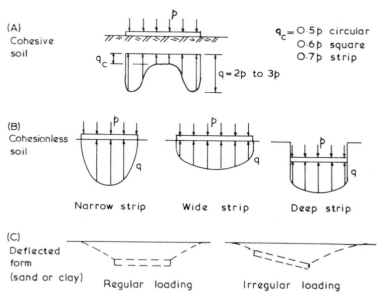

Fig. 3.3

There is therefore ample justification for the assumption of uniform contact pressure values, or uniform variation of values, for most straightforward cases. However, there can be occasions when the loading or the subgrade conditions are such that the designer may modify his calculated values to allow for some of the contact pressure distribution characteristics discussed.

Combined footings

Quite often it is advantageous to provide a single slab foundation to support two, or more, columns. There are two principal conditions which make such a combination necessary:

> Due to property lines there may not be enough space available to provide an exterior column with its own pad foundation. In such

Finite difference numerical techniques for foundations

cases the column must be supported on a foundation which also carries the adjacent interior column

Due to the bearing capacity of the soil it may be that single pad foundations would overlap, or come close together; a row of columns supported on a single slab or beam would be more economical.

Shape of combined footings

Whenever possible combined foundations should have a constant cross section and should be rectangular in plan. However, in order to obtain a uniform bearing pressure beneath the foundation, it is necessary to have the centre of the foundation area coincident with the line of action of the resultant of the applied loads. When this cannot be achieved with a rectangular shaped foundation some other shape must be adopted, usually trapezoidal. Alternatively, a deeper foundation beam can be used.

Design methods for combined footings

As already stated, for the design of a foundation, i.e. the determination of its dimensions and its reinforcement, it is necessary to assess the contact pressures that will act beneath it.

For combined foundations three approaches are open to the designer, each based on different assumptions which are briefly described below.

Rigid method

This is the most common procedure and is often referred to as the 'conventional method'. It is based on two assumptions:

> The foundation is assumed to have perfect rigidity such that its deflection does not affect the contact pressure distribution

> The contact pressure has a straight line distribution such that the line of action of the resultant of the applied loading passes through the centroid of the contact pressure diagram (fig. 3.4A).

Elastic foundation with a simplified elastic subgrade

The rigidity of the foundation is taken as EI (see Appendix II) and the soil is assumed to be replaced by a bed of identical springs,

Elements of Foundation Design

equally spaced and acting independently of each other (fig. 3.4B).

Elastic foundation and subgrade

The rigidity of the foundation is again taken as EI but, this time, the supporting soil is assumed to be fully elastic with Hooke's Law applicable in all directions (fig. 3.4C).

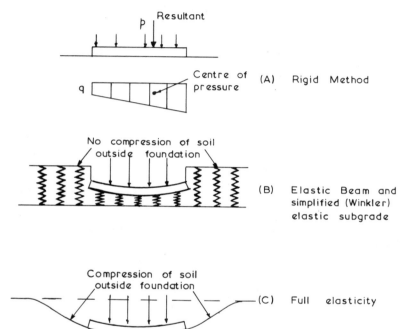

Fig. 3.4 Design assumptions for combined foundations

The rigid method gives a solution with the minimum amount of arithmetic but the other two methods involve the use of numerical techniques. The operation of each method can best be illustrated by working through an example.

Example 3.1

Assume that fig. 3.5 represents a typical foundation beam subjected to the loading shown and that it is required to obtain the bending moment and shear force diagrams for the beam. The subgrade can be assumed to be a stiff clay.

Finite difference numerical techniques for foundations

So that the values obtained from each method can be compared it will be assumed that the beam is of reinforced concrete and has a constant cross section of 500 mm deep by 700 mm wide. The beam carries a uniform load, including its own weight, of 16 kN/m run.

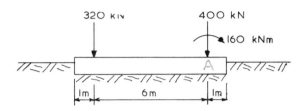

Fig. 3.5 Foundation beam example

(a) Solution by rigid method

The procedure is as follows:
  Determine R_v, the resultant vertical force of the applied loading
  Determine e, the eccentricity of R_v about the centre line of the foundation
  Determine maximum and minimum bearing pressures from standard formulae
  Determine shear force and bending moment values.

$$R_v = 400 + 320 + 16 \times 8 = 848 \text{ kN}.$$

Taking moments about left hand edge of beam:

$$R_v x = 16 \times \frac{8^2}{2} + 320 \times 1 + 400 \times 7 + 160$$

where x = distance of R_v from left hand edge

$$\therefore x = \frac{3792}{848} = 4.472 \text{ m}$$

and R_v acts at e = 0.472 m to the right of the centre line of the foundation.
  The maximum and minimum bearing pressures can be obtained from standard formulae available in most soils text books.

108   *Elements of Foundation Design*

In this case e is within the middle third and

$$q_{min}^{max} = \frac{R_v}{L}\left(1 \pm \frac{6e}{L}\right) \times \frac{1}{B}$$

$$= \frac{848}{8}\left(1 \pm \frac{6 \times 0.472}{8}\right) \times \frac{1}{0.7}$$

$$= 205.02 \text{ and } 97.83 \text{ kN/m}^2.$$

In order to draw the shear force diagram it is best to think of q, not as a pressure, but as a uniformly varying load beneath the foundation.

Maximum upward force = $0.7 \times 205.02 = 143.52$ kN/m.
Minimum upward force = $0.7 \times 97.83 = 68.48$ kN/m.

The 'P and Q' force system, the downward and upward reactive forces at relevant sections along the beam, can now be evaluated and the bending moment and shear force diagrams obtained (fig. 3.6).

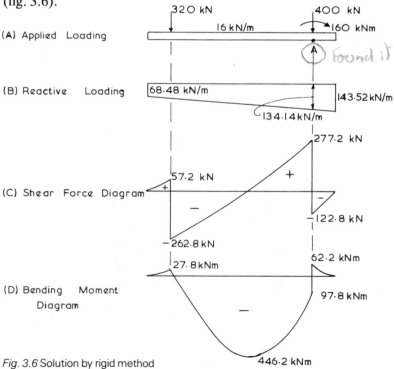

Fig. 3.6 Solution by rigid method

Finite difference numerical techniques for foundations

The calculations are straightforward and only typical ones will be shown. Shear force – convention is:

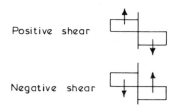

Shear force at point A
 Considering forces to the right of A (i.e. without the 400 kN load)

$$F = -\frac{(143.52 + 134.14)}{2} \times 1.0 + 16 \times 1.0 = -122.83 \text{ kN.}$$

Considering forces to the left of A (i.e. with the 400 kN load)

$$F = 400 - 122.83 = 277.17 \text{ kN.}$$

Bending moment–convention is:

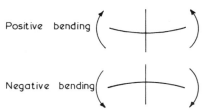

Bending moment at A:
 Considering forces to the right of A (i.e. without the 160 kNm moment)

$$M = 134.1 \times \frac{1}{2} + \frac{9.38}{2} \times \frac{2}{3} - 16 \times \frac{1}{2} = 62.2 \text{ kNm.}$$

Considering the effect of the 160 kNm moment will give

$$M = 62.2 - 160 = -97.8 \text{ kNm.}$$

Various points along the beam can be selected for evaluation of bending moments and the resulting diagram is as shown in fig. 3.6. Maximum bending moment is 446 kNm.

(b) Solution by the simplified elastic subgrade method

This method, often referred to as the Winkler foundation, assumes that the subgrade is replaced by a series of soil springs (fig. 3.4B). The theory was first introduced by Winkler in 1867, has since been extended, and is now in common use in the analysis of foundation problems. As the theory involves the use of a coefficient of subgrade reaction some discussion of this coefficient must take place before attempting to solve the problem.

Coefficient of subgrade reaction

Provided that a spring is not stressed beyond its limit of linearity, its stiffness has a constant value, equal to the spring load divided by the spring extension.

In a similar manner the ratio between the subgrade reactive pressure, q, at a point immediately below a loaded foundation and the settlement of the point, y, can be considered to give a measure of the soil stiffness at the point. This ratio is given the symbol k_s and is known as the coefficient (or modulus) of subgrade reaction. In symbols:

$$k_s = \frac{q}{y}$$

The units for k_s are usually kN/m^3 or MN/m^3.

The assumption is hardly accurate. Even with a perfectly homogeneous soil it is obvious that k_s cannot be a constant throughout the soil mass as uniformly loaded foundations experience non-uniformly varying settlement, unless they are perfectly rigid. However, the method gives realistic values for contact pressures and is suitable for beam design, when only the order of settlement values is required.

Factors affecting the value of k_s.

This subject was discussed by Terzaghi (1955) and only a brief synopsis will be given here.

Consider two foundation beams, of width B_1 and B_2 (such that $B_2 = nB_1$) resting on a compressible subgrade and each loaded so that the subgrade reactive pressure is uniform and equal to q for both beams (fig. 3.7).

Consider the same points on each beam and let y_1 = settlement of beam of width B_1 and let y_2 = settlement of beam of width B_2.

Hence $\quad k_{s_1} = \dfrac{q}{y_1} \quad$ and $\quad k_{s_2} = \dfrac{q}{y_2}$.

Finite difference numerical techniques for foundations 111

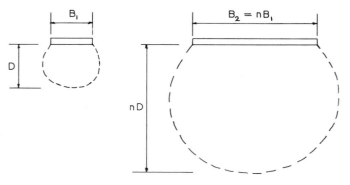

Fig. 3.7 Bulbs of pressure for vertical stress for different beams

If the beams are resting on a subgrade whose deformation properties are more or less independent of depth (such as a stiff clay) then it can be assumed that the settlement increases in simple proportion to the depth of the pressure bulb.

Then $\quad y_2 = ny_1$

and $\quad k_{s_2} = \dfrac{q}{ny_1} = \dfrac{q}{y_1} \cdot \dfrac{B_1}{B_2} = k_{s_1}\dfrac{B_1}{B_2}.$

A general expression for k_s can now be obtained if we consider B_1 as being of unit length. (Terzaghi used a unit length of one foot, which must now be converted into a metric measurement = 0.305 m).

Hence by putting 0.305 m for B_1; k_s for k_{s_2}; B for B_2 we obtain the expression:

$$k_s = 0.305 \dfrac{k_{s_1}}{B}$$

where k_s is the coefficient of subgrade reaction of a foundation of width B m and resting on a stiff clay.

k_{s_1} is the coefficient of subgrade reaction of a foundation of width 0.305 m, resting on the same clay. Note that the value of k_{s_1} is derived from ultimate settlement values, i.e. after consolidation settlement has occurred.

If the beams are resting on clean sand the final settlement values are obtained almost instantaneously but, as the modulus of elasticity of sand increases with depth, the deformation characteristics of the sand change and become less compressible with depth. Because of this characteristic of sand the lower portion of the bulb of pressure for beam B_2 is less compressible than that of the sand enclosed in the bulb of pressure for beam B_1.

The settlement value y_2 has therefore a value somewhere between y_1 and ny_1. It has been shown experimentally (Terzaghi and Peck, 1948) that the settlement, y, of a beam of width B m, resting on sand is given by the expression:

$$y = y_1 \left(\frac{2B}{B + 0.305} \right)^2$$

where y_1 = settlement of a beam of width 0.305 m subjected to the same subgrade reactive pressure as the beam of width B m.

Hence, the coefficient of subgrade reaction, k_s, of a beam of width B m on sand is:

$$k_s = \frac{q}{y} = \frac{q}{y_1} \left(\frac{B + 0.305}{2B} \right)^2 = k_{s_1} \left(\frac{B + 0.305}{2B} \right)^2$$

where k_{s_1} = coefficient of subgrade reaction of a beam of width 0.305 m resting on the same sand.

In summary, k_s for a beam of width B m can be obtained from the following formulae:

$$k_s = 0.305 \frac{k_{s_1}}{B} \quad \text{for stiff clays}$$

$$k_s = k_{s_1} \left(\frac{B + 0.305}{2B} \right)^2 \quad \text{for sands.}$$

Measurement of a value for k_{s_1}

A value for k_{s_1} for a particular subgrade can be obtained by carrying out plate loading tests and averaging the results as described in chapter 1. When the modulus of subgrade reaction is determined with the use of 0.305 m square plates it is given a special symbol, $\bar{k}_{s_1}$.

For sands $k_{s_1} \approx \bar{k}_{s_1}$ but for clays k_{s_1} varies with the length of the beam. Terzaghi (1955) gives the formula (for clays)

$$k_{s_1} = \bar{k}_{s_1} \left(\frac{\ell + 0.152}{1.5 \ell} \right)$$

where ℓ is the length of the beam in metres and the width of the beam is 0.305 m. In the limit, for an infinitely long beam sitting on clay,

$$k_{s_1} = \frac{\bar{k}_{s_1}}{1.5}.$$

The procedure necessary to find the k_s value to be used in a beam design is therefore:

Determine length of foundation beam.
Given $\bar{k}_{s_1}$ (from a test or an estimation) determine k_{s_1} using Terzaghi's formulae:

For sands $k_{s_1} = \bar{k}_{s_1}$; for clays $k_{s_1} = \bar{k}_{s_1} \left(\dfrac{\ell + 0.152}{1.5\,\ell} \right)$.

Knowing k_{s_1}, determine k_s as described previously.

Example. A foundation beam resting on clay. $\bar{k}_{s_1} = 25$ MN/m^3.

Length of beam = 4 m.   Width of beam = 0.75 m

$$k_{s_1} = 25\,\frac{(4 + 0.152)}{1.5 \times 4} = 17.3 \text{ MN/m}^3$$

$$\therefore k_s = 0.305 \times \frac{17.3}{0.75} = 7 \text{ MN/m}^3.$$

Typical test results are shown in fig. 3.8. The Road Research Laboratory (1952) recommend that $\bar{k}_{s_1}$ is determined by taking the bearing pressure corresponding to a settlement of 1.3 mm. In fig. 3.8 this gives a $\bar{k}_{s_1}$ value of 34 MN/m^3.

It should be noted that for cohesionless soils the settlements of the plate are produced almost instantaneously with the application of each loading increment. With clay the settlement increases with time so that the final value determined for $\bar{k}_{s_1}$ will depend upon the length of time of the test. Obviously increments should be applied slowly enough to allow full settlement values to be reached. It is these final settlement values that are used to determine $\bar{k}_{s_1}$.

Estimation of $\bar{k}_{s_1}$ values

Plate loading tests are both costly and time consuming and are often not necessary, unless the foundation designer is interested in predicting accurate settlement values.

114 *Elements of Foundation Design*

Generally a designer only requires the values of the bending moments and the shear forces within his foundation. With even a relatively large error in the estimation of k_s, moments and shear forces can be calculated with little error (Terzaghi, 1955); an error of 100% in the estimation of k_s may change the structural behaviour of the foundation by up to 15% only.

In the absence of plate loading tests estimated values for $\bar{k}_{s_1}$, and hence k_s, are used. The values suggested by Terzaghi, converted into S.I. units and rounded off, are set out below:

Cohesionless (or slightly cohesive) soils

Relative density	Loose	Medium	Dense
Standard penetration test uncorrected N value	< 10	10–30	> 30
$\bar{k}_{s_1}$ (MN/m³) soil dry or moist	15	45	175
$\bar{k}_{s_1}$ (MN/m³) soil submerged	10	30	100

Stiff clays

Consistency	Stiff	Very stiff	Hard
$\underline{c}_u$ (kN/m²)	105–215	215–430	> 430
$\bar{k}_{s_1}$ (MN/m³)	25	50	100

Note. For soft normally consolidated clays $\bar{k}_{s_1}$ values are so small that they cannot be meaningfully used and the designer is forced to use the rigid method previously described.

Note. The Winkler assumption that q/y equals a constant gives the straight line OA in fig. 3.8. It is seen that this assumption is realistic up to bearing pressure values of about ½ × the ultimate bearing capacity of the subgrade. The Winkler assumption should therefore only be used for contact pressures of less than half the ultimate bearing capacity of the soil.

The actual solution of example 3.1 by the simplified elastic subgrade method can now be attempted.

The problem is illustrated in fig. 3.5 and the dimensions to be used are:

Depth of beam = 500 mm
Width of beam = 700 mm

Finite difference numerical techniques for foundations

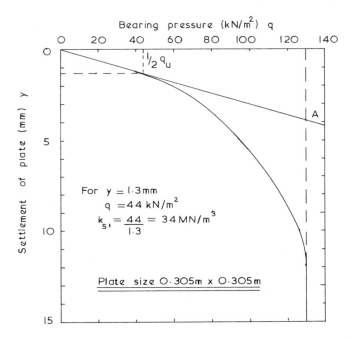

Fig. 3.8 Typical plate load test results

The following properties will be assumed:

Modulus of elasticity for concrete, E_c = 20 000 MN/m²
Modulus of subgrade reaction, k_{s_1} = 30 MN/m³

Note. The numerical method to be used here is that of finite differences, explained in Appendix III. For reasonably accurate results the beam should be divided into at least eight equal sections, to give nine nodal points; this is within the capabilities of some of the programmable calculators now available.

Before the eight section analysis is carried out, for illustrative purposes, the beam will be divided into four equal sections, i.e. a = 2 m. It will be seen that, even with only five nodal points the values obtained for the contact pressure distribution would be sufficient for most practical design problems.

The division of the beam into four equal sections is snown in fig. 3.9A.

116 *Elements of Foundation Design*

There are five nodal points; let their deflections be y_1, y_2, y_3, y_4 and y_5.

Then $q_1 = k_s y_1$; $q_2 = k_s y_2 \ldots$ etc.

For a stiff clay $k_s = \dfrac{k_{s1}}{B} \times 0.305 = \dfrac{30}{0.7} \times 0.305 = 13 \text{ MN/m}^3$.

Moment of inertia of beam $= \dfrac{1}{12} \times BD^3$.

$\therefore EI = 20\,000 \times \dfrac{1}{12} \times 0.7 \times (0.5)^3 = 145.75 \text{ MNm}^2$.

Assume that the contact pressures act as concentrated upward reactive forces, Q_1, Q_2, Q_3, Q_4 and Q_5 at the nodal points 1, 2, 3, 4 and 5.

$$Q_1 = q_1 \times \text{area} = q_1 \times \dfrac{a}{2} \times B = k_s y_1 \dfrac{a}{2} B$$

$$= 13 \times \dfrac{2}{2} \times 0.7 \times y_1 = 9.1 \, y_1.$$

Similarly, $Q_2 = 18.2 \, y_2 \ldots$ etc.

There is no applied moment at either end of the beam so it can be assumed that points 1 and 5 are not fixed and that $M_1 = M_5 = 0$.

Expressions for M_2, M_3 and M_4 can be evaluated:

$$M_2 = Q_1 \times 2 - 320 \times 1 - 16 \times \dfrac{2^2}{2} = 18.2 \, y_1 - 352 \quad (1)$$

$$M_3 = 36.4 \, y_1 + 36.4 \, y_2 - 1088 \quad (2)$$

$$M_4 = 54.6 \, y_1 + 72.8 \, y_2 + 36.4 \, y_3 - 1888. \quad (3)$$

From Appendix II we know that $- M = EI \dfrac{d^2 y}{dx^2}$

and, from Appendix III: $\dfrac{d^2 y}{dx^2} = \dfrac{y_1 - 2y_2 + y_3}{a^2}$

hence, equation 1 may be rewritten as:

$$- 18.2 \, y_1 + 352 = \dfrac{145.750}{4} (y_1 - 2y_2 + y_3)$$

i.e
$$1.5\,y_1 - 2y_2 + y_3 = 9.66. \qquad (1)$$

Similarly for equations 2 and 3:

$$1.0\,y_1 + 2.0\,y_2 - 2y_3 + y_4 = 29.86 \qquad (2)$$

$$1.5\,y_1 + 2.0\,y_2 + 2.0\,y_3 - 2.0\,y_4 + 1.0\,y_5 = 51.81. \qquad (3)$$

(Note that EI is expressed as 145.750 MNm^2 so that the y values will be in millimetres).

There are five unknowns (y_1 to y_5) so two further simultaneous equations are required. These can be obtained by equating M_1 (or M_5) to zero and equating ΣR to zero.

$\Sigma M_5 = 0$ gives:

$$72.8\,y_1 + 109.2\,y_2 + 72.8\,y_3 + 36.4\,y_4 = 2992 \qquad (4)$$

$$9.1\,y_1 + 18.2\,y_2 + 18.2\,y_3 + 18.2\,y_4 + 9.1\,y_5 = 848 \qquad (5)$$

Expressing in matrix form:

$$\begin{bmatrix} 1.5 & -2 & 1 & 0 & 0 \\ 1.0 & 2 & -2 & 1 & 0 \\ 1.5 & 2 & 2 & -2 & 1 \\ 72.8 & 109.2 & 72.8 & 36.4 & 0 \\ 9.1 & 18.2 & 18.2 & 18.2 & 9.1 \end{bmatrix} \begin{bmatrix} y_1 \\ y_2 \\ y_3 \\ y_4 \\ y_5 \end{bmatrix} = \begin{bmatrix} 9.66 \\ 29.86 \\ 51.81 \\ 2992.0 \\ 848.0 \end{bmatrix}$$

By inverting the matrix we obtain:

$$\begin{bmatrix} 0.3687 & 0.2581 & 0.0829 & 0.0020 & -0.0091 \\ -0.2396 & 0.0323 & 0.0461 & 0.0042 & -0.0051 \\ -0.0323 & -0.3226 & -0.0323 & 0.0053 & 0.0035 \\ 0.0461 & 0.0323 & -0.2396 & 0.0003 & 0.0263 \\ 0.0829 & 0.2581 & 0.3687 & -0.0215 & 0.0694 \end{bmatrix} \begin{bmatrix} 9.66 \\ 29.86 \\ 51.81 \\ 2992 \\ 848 \end{bmatrix} = \begin{bmatrix} y_1 \\ y_2 \\ y_3 \\ y_4 \\ y_5 \end{bmatrix}$$

Note. To reduce rounding off errors it is important to work to at least four decimal places in the inverse matrix.

Multiplying out leads to:

Deflection (mm)	Q values (kN)	q values (kN/m^2)	Corresponding reactive forces/m length of beam
$y_1 = 13.8$	$Q_1 = 125.9$	$q_1 = 179.8$	125.9
$y_2 = 9.3$	$Q_2 = 168.9$	$q_2 = 120.6$	84.4
$y_3 = 7.2$	$Q_3 = 131.2$	$q_3 = 93.7$	65.6
$y_4 = 12.2$	$Q_4 = 222.0$	$q_4 = 158.6$	111.0
$y_5 = 22.1$	$Q_5 = 201.4$	$q_5 = 287.7$	201.4

Check $\Sigma = 849.4$ (Round off error = 1.4)

118 *Elements of Foundation Design*

Evaluation of the Q forces affords a useful check on the accuracy of the computation. In order to obtain the bending moment and shear force diagrams it is best to work from the contact pressure values, q, which are simply equal to k_s y (in kN/m^2). As the beam is only 0.7 m wide it is best to modify the contact pressure diagram by multiplying the value of q by 0.7 to give a disturbed reaction in kN/m length of beam. Once this has been done the completed loading diagram is as shown in fig. 3.9A.

Bearing in mind the form of applied loading, fig. 3.9A looks considerably more convincing than fig. 3.6A.

It will be seen that the simplified elastic subgrade effort leads to a higher value of maximum bearing pressure and a considerable reduction in maximum bending moment in the foundation.

Improvement in accuracy

For greater accuracy the beam should be split into at least eight equal sections. When this is done

$$Q_1 = \frac{B}{2} k_{s_1} y_1 = 4.55 \, y_1$$

and $Q_2 = 9.1 \, y_2$ ... etc (fig. 3.10A). The expressions for the bending moments M_2 to M_8 are:

$M_2 = 4.55 \, y_1 - 8$
$M_3 = 9.1 \, y_1 + 9.1 \, y_2 - 352$
$M_4 = 13.65 \, y_1 + 18.2 \, y_2 + 9.1 \, y_3 - 712$
$M_5 = 18.2 \, y_1 + 27.3 \, y_2 + 18.2 \, y_3 + 9.1 \, y_4 - 1088$
$M_6 = 22.75 \, y_1 + 36.4 \, y_2 + 27.3 \, y_3 + 18.2 \, y_4 + 9.1 \, y_5 - 1480$
$M_7 = 27.3 \, y_1 + 45.5 \, y_2 + 36.4 \, y_3 + 27.3 \, y_4 + 18.2 \, y_5 + 9.1 \, y_6 - 1888$
$M_8 = 31.85 \, y_1 + 54.6 \, y_2 + 45.5 \, y_3 + 36.4 \, y_4 + 27.3 \, y_5 + 18.2 \, y_6 + 9.1 \, y_7 - 2232.$

Note, M_8 is taken to be the average of the two moment values that occur at point 8 when the applied moment of 160 kNm is allowed for, i.e. when calculating M_8 either take moments of all the forces to the left of point 8 and add 160/2 kNm or take moments of all forces to the right of point 8 and subtract 160/2 kNm.

(A) Loading

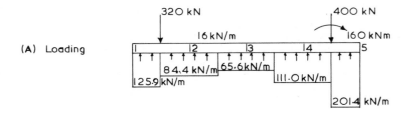

(B) Shear Force Diagram

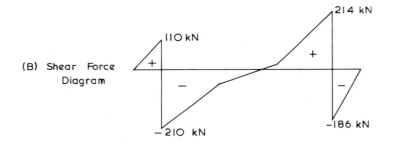

(C) Bending Moment Diagram

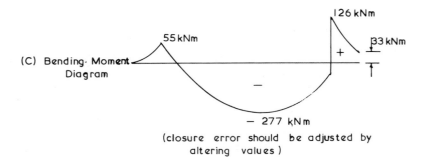

(closure error should be adjusted by altering values)

Fig. 3.9 Solution by simplified elastic subgrade method using four slices

The remaining two required simultaneous equations are obtained from $M_9 = 0$ and $\Sigma R = 0$. The final matrix form of the equations is:

$$\begin{bmatrix} 1.0312 & -2.0 & 1.0 & 0 & 0 & 0 & 0 & 0 & 0 \\ 0.0624 & 1.0624 & -2.0 & 1.0 & 0 & 0 & 0 & 0 & 0 \\ 0.0937 & 0.1249 & 1.0624 & -2.0 & 1.0 & 0 & 0 & 0 & 0 \\ 0.1249 & 0.1873 & 0.1249 & 1.0624 & -2.0 & 1.0 & 0 & 0 & 0 \\ 0.1561 & 0.2497 & 0.1873 & 0.1249 & 1.0624 & -2.0 & 1.0 & 0 & 0 \\ 0.1873 & 0.3122 & 0.2497 & 0.1873 & 0.1249 & 1.0624 & -2.0 & 1.0 & 0 \\ 0.2185 & 0.3746 & 0.3122 & 0.2497 & 0.1873 & 0.1249 & 1.0624 & -2.0 & 1.0 \\ 36.4 & 63.7 & 54.6 & 45.5 & 36.4 & 27.3 & 18.2 & 9.1 & 0 \\ 4.55 & 9.1 & 9.1 & 9.1 & 9.1 & 9.1 & 9.1 & 9.1 & 4.55 \end{bmatrix} \begin{bmatrix} y_1 \\ y_2 \\ y_3 \\ y_4 \\ y_5 \\ y_6 \\ y_7 \\ y_8 \\ y_9 \end{bmatrix} = \begin{bmatrix} 0.0549 \\ 2.4151 \\ 4.8851 \\ 7.4648 \\ 10.1544 \\ 12.9537 \\ 15.8628 \\ 2992.0 \\ 848.0 \end{bmatrix}$$

The inversion of the matrix leads to:

$$\begin{bmatrix} 0.6533 & 0.8486 & 0.7806 & 0.5909 & 0.3724 & 0.1812 & 0.0499 & 0.0020 & -0.0110 \\ -0.2290 & 0.1784 & 0.3310 & 0.3263 & 0.2412 & 0.1322 & 0.0407 & 0.0034 & -0.0089 \\ -0.1317 & -0.5182 & -0.1429 & 0.0433 & 0.0983 & 0.0775 & 0.0300 & 0.0047 & -0.0066 \\ -0.0608 & -0.2790 & -0.6862 & -0.2970 & -0.0828 & 0.0033 & 0.0136 & 0.0056 & -0.0030 \\ -0.0144 & -0.1092 & -0.3351 & -0.7361 & -0.3351 & -0.1092 & -0.0144 & 0.0057 & 0.0032 \\ 0.0136 & 0.0033 & -0.0828 & -0.2970 & -0.6863 & -0.2790 & -0.0608 & 0.0039 & 0.0134 \\ 0.0300 & 0.0775 & 0.0433 & -0.0433 & -0.1429 & -0.5182 & -0.1317 & -0.0009 & 0.0289 \\ 0.0407 & 0.1322 & 0.2412 & 0.3263 & 0.3310 & 0.1785 & -0.2290 & -0.0104 & 0.0503 \\ 0.0499 & 0.1811 & 0.3724 & 0.5909 & 0.7806 & 0.8486 & 0.6533 & -0.0259 & 0.0762 \end{bmatrix} \begin{bmatrix} 0.0549 \\ 2.4151 \\ 4.8851 \\ 7.4648 \\ 10.1544 \\ 12.9537 \\ 15.8628 \\ 2992 \\ 848 \end{bmatrix} = \begin{bmatrix} y_1 \\ y_2 \\ y_3 \\ y_4 \\ y_5 \\ y_6 \\ y_7 \\ y_8 \\ y_9 \end{bmatrix}$$

from which the bending moment and shear force diagrams may be obtained (fig. 3.10).

Note. The calculations should be worked through using four figures of decimals. This involves little extra effort when using a calculator and considerably reduces the closure errors in the bending moment and shear force diagrams. After the final values have been evaluated they can then be rounded off.

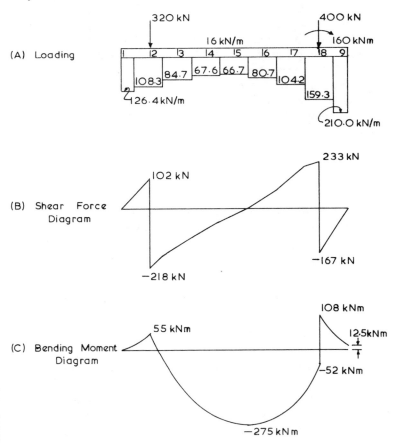

Fig. 3.10 Solution by simplified elastic subgrade method using eight slices

(c) Solution by the elastic foundation and subgrade method
The assumption of full elasticity of the subgrade is considered by

122 Elements of Foundation Design

some to be more realistic than that of Winkler. Although reasonable for rock and saturated clay the assumption is very approximate for other soil types.

There is also the problem of determining an accurate value for E_s, the modulus of elasticity of the soil and μ, Poisson's ratio, which can vary from 0.1 for a wet clay to 0.5 for a saturated clay, with a wide range of values between these two limits for other soil types.

The method has been fully described by Cheung (1977).

In spite of the claims of its devotees, there does not seem much point, in practical problems, to use this method in preference to the Winkler assumption. However, for the sake of completeness, a simplified approach is set out below.

Consider a concentrated load, P, acting on the surface of a semi-infinite elastic medium.

According to Boussinesq (1885) the vertical displacements at the surface of the medium are given by the expression:

$$y = \frac{P(1 - \mu^2)}{\pi r E_s}$$

where μ = Poisson's ratio
E_s = modulus of elasticity of the subgrade
r = radial distance from P to the point considered (fig. 3.11).

The above expression is often written in the form:

$$y = \frac{JP}{E_s r}$$

where J is a dimensionless number equal to $\frac{1 - \mu^2}{\pi}$

(In the light of the remarks regarding the variation of μ it is seen that J can vary from 0.3151 ($\mu = 0.1$) to 0.2387 ($\mu = 0.5$).)

The concept of a concentrated load is a theoretical one and it can be seen that the equation breaks down when $r = 0$. However, if it is assumed that the load P is applied to the soil through a square plate (of dimensions B × B) where B is the width of the beam to be analysed, then it can be shown that the deflection of the surface immediately below P, i.e. y_P, is given by the expression:

$$y_P = \frac{3PJ}{E_s B}$$

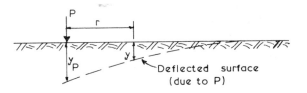

Fig. 3.11

By the principle of superposition it is possible to determine surface deformations below a loaded area if the area can be divided into suitable sections and the load carried by each section assumed to be concentrated at its centroid.

In dealing with beam design here the concentrated load to be considered is Q, the subgrade reactive force and the equations must be rewritten as:

$$y = \frac{QJ}{E_s r} \quad \text{and} \quad y_P = \frac{3QJ}{E_s B}.$$

The analysis of the example will be carried out by finite differences so that the beam must be divided into a series of equal sections, of length a, which will not generally be the same value as the width of the beam, B.

Hence the expressions, when written in matrix form, will be:

$$\begin{bmatrix} y_1 \\ y_2 \\ y_3 \\ \cdot \\ \cdot \\ \cdot \\ y_n \end{bmatrix} = \frac{J}{BE_s} \begin{bmatrix} 3 & 1 \times \frac{B}{a} & \frac{1}{2} \times \frac{B}{a} & \frac{1}{3} \times \frac{B}{a} & \frac{1}{4} \times \frac{B}{a} & \cdot & \cdot \\ 1 \times \frac{B}{a} & 3 & 1 \times \frac{B}{a} & \frac{1}{2} \times \frac{B}{a} & \frac{1}{3} \times \frac{B}{a} & \cdot & \cdot \\ \frac{1}{2} \times \frac{B}{a} & 1 \times \frac{B}{a} & 3 & 1 \times \frac{B}{a} & \frac{1}{2} \times \frac{B}{a} & \cdot & \cdot \\ \cdot & \cdot & \cdot & \cdot & \cdot & \cdot & \cdot \\ \cdot & \cdot & \cdot & \cdot & \cdot & \cdot & \cdot \\ \cdot & \cdot & \cdot & \cdot & \cdot & \cdot & \cdot \\ \cdot & \cdot & \cdot & \cdot & \cdot & \cdot & \cdot \end{bmatrix} \begin{bmatrix} Q_1 \\ Q_2 \\ Q_3 \\ \cdot \\ \cdot \\ \cdot \\ Q_n \end{bmatrix}$$

Elements of Foundation Design

Explanation

Assume that the beam is divided into four equal sections (fig. 3.12). Consider the effect of Q_1

Settlement at point 1, due to $Q_1 = \dfrac{3Q_1 J}{E_s B}$

Settlement at point 2, due to $Q_1 = \dfrac{Q_1 J}{E_s r}$

where r = distance from $Q_1 = a = a \times \dfrac{B}{B}$.

Hence, settlement at point 2, due to $Q_1 = \dfrac{Q_1 J}{E_s B} \times 1 \times \dfrac{B}{a}$.

Settlement at point 3, due to Q_1

$= \dfrac{Q_1 J}{E_s r} = \dfrac{Q_1 J}{E_s 2a} = \dfrac{Q_1 J}{E_s B} \times \dfrac{1}{2} \times \dfrac{B}{a}$.

Hence, for y_1, the row of the matrix becomes:

$$\left[3 \quad 1 \times \dfrac{B}{a} \quad \dfrac{1}{2} \times \dfrac{B}{a} \quad \dfrac{1}{3} \times \dfrac{B}{a} \quad \ldots \right]$$

and the matrix is multiplied by the constant $\dfrac{J}{BE_s}$.

By considering the loading diagram shown in fig. 3.12 expressions for the bending moments at each of the nodal points can be readily obtained:

$$M_1 = 0$$
$$M_2 = 2Q_1 - 352$$
$$M_3 = 4Q_1 + 2Q_2 - 1088$$
$$M_4 = 6Q_1 + 4Q_2 + 2Q_3 - 1888$$
$$M_5 = 0.$$

Now

$$M = -EI \dfrac{d^2 y}{dx^2}$$

which can be expressed in finite difference form (see Appendix III).

Finite difference numerical techniques for foundations

$$\begin{bmatrix} M_1 \\ M_2 \\ M_3 \\ M_4 \\ M_5 \end{bmatrix} = -\frac{EI}{a^2} \begin{bmatrix} -2 & 1 & 0 & 0 & 0 \\ 1 & -2 & 1 & 0 & 0 \\ 0 & 1 & -2 & 1 & 0 \\ 0 & 0 & 1 & -2 & 1 \\ 0 & 0 & 0 & 1 & -2 \end{bmatrix} \begin{bmatrix} y_1 \\ y_2 \\ y_3 \\ y_4 \\ y_5 \end{bmatrix}$$

$$\frac{EI}{a^2} = \frac{145.75}{4} \text{ (with y values in millimetres).}$$

M_1 and M_5 are both equal to zero and can be removed from the above equation provided that the top and bottom rows of the matrix are also removed.

The expression then becomes:

$$\begin{bmatrix} M_2 \\ M_3 \\ M_4 \end{bmatrix} = -\frac{145.75}{4} \begin{bmatrix} 1 & -2 & 1 & 0 & 0 \\ 0 & 1 & -2 & 1 & 0 \\ 0 & 0 & 1 & -2 & 1 \end{bmatrix} \begin{bmatrix} y_1 \\ y_2 \\ y_3 \\ y_4 \\ y_5 \end{bmatrix} \quad (A)$$

Now, assuming $E_s = 20$ MN/m² and $\mu = 0.5$, whence $J = 0.2387$ and putting $B = 0.7$ m and $a = 2$ m, the matrix expression relating the vector y to the vector Q becomes:

$$\begin{bmatrix} y_1 \\ y_2 \\ y_3 \\ y_4 \\ y_5 \end{bmatrix} = \frac{0.2387}{20\,000 \times 0.7} \begin{bmatrix} 3 & 0.35 & 0.175 & 0.1167 & 0.0875 \\ 0.35 & 3 & 0.35 & 0.175 & 0.1167 \\ 0.175 & 0.35 & 3 & 0.35 & 0.175 \\ 0.1167 & 0.175 & 0.35 & 3 & 0.35 \\ 0.0875 & 0.1167 & 0.175 & 0.35 & 3 \end{bmatrix} \begin{bmatrix} Q_1 \\ Q_2 \\ Q_3 \\ Q_4 \\ Q_5 \end{bmatrix}$$

Hence, substituting for y in equation (A):

$$\begin{bmatrix} M_2 \\ M_3 \\ M_4 \end{bmatrix} = -\frac{0.2387 \times 145.75}{20 \times 0.7 \times 4} \begin{bmatrix} 1 & -2 & 1 & 0 & 0 \\ 0 & 1 & -2 & 1 & 0 \\ 0 & 0 & 1 & -2 & 1 \end{bmatrix} \begin{bmatrix} 3 & 0.35 & 0.175 & 0.1167 & 0.0875 \\ 0.35 & 3 & 0.35 & 0.175 & 0.1167 \\ 0.175 & 0.35 & 3 & 0.35 & 0.175 \\ 0.1167 & 0.175 & 0.35 & 3 & 0.35 \\ 0.0875 & 0.1167 & 0.175 & 0.35 & 3 \end{bmatrix} \begin{bmatrix} Q_1 \\ Q_2 \\ Q_3 \\ Q_4 \\ Q_5 \end{bmatrix}$$

Multiplying out this latest expression gives:

$$\begin{bmatrix} M_2 \\ M_3 \\ M_4 \end{bmatrix} = -0.6213 \begin{bmatrix} 2.475 & -5.3 & 2.475 & 0.1167 & 0.0291 \\ 0.1167 & 2.475 & -5.3 & 2.475 & 0.1167 \\ 0.0291 & 0.1167 & 2.475 & -5.3 & 2.475 \end{bmatrix} \begin{bmatrix} Q_1 \\ Q_2 \\ Q_3 \\ Q_4 \\ Q_5 \end{bmatrix}$$

Substituting the expressions for M_2, M_3 and M_4 three equations relating the five Q values, are obtained. The remaining required two equations are obtained from $M_5 = 0$ and $\Sigma R = 0$ and the final expression is:

$$\begin{bmatrix} 5.694 & -5.3 & 2.475 & 0.1167 & 0.0291 \\ 6.5547 & 5.694 & -5.3 & 2.475 & 0.1167 \\ 9.686 & 6.5547 & 5.694 & -5.3 & 2.475 \\ 8 & 6 & 4 & 2 & 0 \\ 1 & 1 & 1 & 1 & 1 \end{bmatrix} \begin{bmatrix} Q_1 \\ Q_2 \\ Q_3 \\ Q_4 \\ Q_5 \end{bmatrix} = \begin{bmatrix} 566.6 \\ 1751.2 \\ 3038.8 \\ 2992 \\ 848 \end{bmatrix}$$

Finite difference numerical techniques for foundations

By inverting the matrix this becomes:

$$\begin{bmatrix} Q_1 \\ Q_2 \\ Q_3 \\ Q_4 \\ Q_5 \end{bmatrix} = \begin{bmatrix} 0.0798 & 0.0535 & 0.0192 & 0.0081 & -0.0561 \\ -0.1062 & 0.0041 & 0.0150 & 0.0583 & -0.0346 \\ -0.0079 & -0.1151 & -0.0079 & 0.1053 & 0.0332 \\ 0.0150 & 0.0041 & -0.1062 & 0.0818 & 0.2619 \\ 0.0192 & 0.0535 & 0.0798 & -0.2536 & 0.7956 \end{bmatrix} \begin{bmatrix} 566.6 \\ 1751.2 \\ 3038.8 \\ 2992 \\ 848 \end{bmatrix}$$

Note. A practical engineer might be a little suspicious of the false accuracy that seems to occur in the above expression. Surely 1751.2 is the same as 1751.0! Why not round off the figures?

The immediate answer must be that he is right. In the days of the slide rule such a procedure was necessary. However, the figures shown above are as the calculator produced them and as it is just as quick to punch out a five figure number as a four why change them? If the process of rounding off is used with a calculator a lot of valuable time can be wasted. The time to round off is when one has the final values. Obviously a settlement of 13.3479 mm must be quoted as 13 mm.

On multiplying out, the reactive forces at the nodal points are found to be: $Q_1 = 173.9$; $Q_2 = 137.7$; $Q_3 = 113.2$; $Q_4 = 159.8$; $Q_5 = 263.0$ kN.

The next step would be to calculate the shear force and bending moment values. As this procedure has already been illustrated it will not be repeated.

As for solution (b), accuracy is increased if the beam is divided into a larger number of sections.

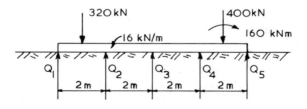

Fig. 3.12 Solution by assumption of full elasticity

Treatment of a foundation of varying width

Generally when using the Winkler foundation, the dimensions of the foundation slab are first determined by a rigid method analysis. There are occasions when the most suitable form of a foundation

128 Elements of Foundation Design

slab is one of constant depth but varying width, the most common type being one which is trapezoidal in plan.

The Winkler analysis can still be carried out as described previously except that the moment of inertia of the foundation slab must be determined for each nodal point and the resulting set of differing EI values inserted into the calculations.

Example 3.2

Details of a proposed reinforced concrete foundation slab are given in fig. 3.13. The slab is of constant 125 mm thickness and the subgrade consists of sand with $\bar{k}_{s_1} = 15 \text{ MN/m}^3$. $E_c = 20\,000 \text{ MN/m}^2$.

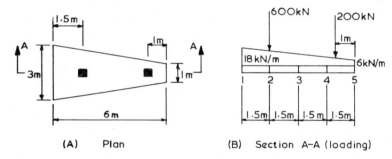

Fig. 3.13 Foundation of varying width

The loading diagram, along centre line of foundation, is as shown in fig. 3.13B.

For sands $\bar{k}_{s_1} = k_{s_1}$.

Average width of foundation, B (for determination of k_s) = 2.0 m

$$k_s = 15 \left(\frac{2 + 0.305}{2 \times 2}\right)^2 = 5.0 \text{ MN/m}^3.$$

Divide slab into four equal sections, to give 5 nodal points (see fig. 3.13)

Nodal point	Width (m)	I (m⁴)	EI (kNm²)
1	3	0.000 4883	9765
2	2.5	0.000 4069	8138
3	2.0	0.000 3255	6510
4	1.5	0.000 2441	4883
5	1	0.000 1628	3255

Finite difference numerical techniques for foundations 129

By considering fig. 3.13B, the reactive forces, Q, at the nodal points are given by:

$$Q_1 = k_s y_1 \times \text{area} = 5 \cdot \times 2.875 \times \frac{1.5}{2} = 10.781\, y_1$$

$$Q_2 = 5 \times 2.5 \times 1.5 \times y_2 \qquad = 18.75\, y_2$$
$$Q_3 \qquad\qquad\qquad\qquad\qquad = 15.0\, y_3$$
$$Q_4 \qquad\qquad\qquad\qquad\qquad = 11.25\, y_4$$
$$Q_5 \qquad\qquad\qquad\qquad\qquad = 4.219\, y_5$$

From fig. 3.13B the expressions for the moments at the nodal points are:

$$M_2 = Q_1 \times 1.5 - 15 \times \frac{1.5^2}{2} - 3 \times \frac{1.5^2}{3} = 1.5\, Q_1 - 19.125$$

$$M_3 = Q_1 \times 3 + Q_2 \times 1.5 - 600 \times 1.5 - 12 \times \frac{3^2}{2} - 6 \times 3$$
$$= 3\, Q_1 + 1.5\, Q_2 - 972$$

$$M_4 = Q_1 \times 4.5 + Q_2 \times 3.0 + Q_3 \times 1.5 - 600 \times 3.0 - 9 \times \frac{4.5^2}{2}$$
$$- 9 \times \frac{4.5^2}{3}$$
$$= 4.5\, Q_1 + 3\, Q_2 + 1.5\, Q_3 - 1951.875$$

$$M_5 = 6\, Q_1 + 4.5\, Q_2 + 3\, Q_3 + 1.5\, Q_4 - 3152.$$

Now $- M_2 = \dfrac{EI}{a^2}(y_1 - 2\, y_2 + y_3)\ldots$etc.

$$\therefore M_2 = \frac{1.5^2}{8138}(16.175\, y_1 - 19.125) = -(y_1 - 2\, y_2 + y_3)$$

i.e. $\qquad 1.00447\, y_1 - 2\, y_2 + y_3 = 0.005287 \qquad\qquad (1)$

$$M_3 = \frac{1.5^2}{6510}(32.343\, y_1 + 28.125\, y_2 - 972)$$
$$= -(y_2 - 2\, y_3 + y_4)$$

i.e. $\qquad 0.011\,178\, y_1 + 1.009\,721\, y_2 - 2\, y_3 + y_4 = 0.3359. \qquad (2)$

Similarly (for M_4):

$$0.022\,35\, y_1 + 0.025\,92\, y_2 + 1.010\,367\, y_3 - 2\, y_4 + y_5 = 0.899\,389.$$
$$\qquad\qquad\qquad\qquad\qquad\qquad\qquad\qquad\qquad (3)$$

130 Elements of Foundation Design

The remaining two equations can be obtained from the facts that $M_5 = 0$ and $\Sigma R = 0$:

$$64.686\, y_1 + 84.375\, y_2 + 45\, y_3 + 16.875\, y_4 = 3152 \qquad (4)$$

$$10.871\, y_1 + 18.75\, y_2 + 15\, y_3 + 11.25\, y_4 + 4.219\, y_5 = 872. \qquad (5)$$

Expressing equations (1) to (5) in matrix form gives:

$$\begin{bmatrix} 1.0045 & -2 & 1 & 0 & 0 \\ 0.0112 & 1.0097 & -2 & 1 & 0 \\ 0.0224 & 0.0259 & 1.0104 & -2 & 1 \\ 64.686 & 84.375 & 45 & 16.875 & 0 \\ 10.781 & 18.75 & 15 & 11.25 & 4.219 \end{bmatrix} \begin{bmatrix} y_1 \\ y_2 \\ y_3 \\ y_4 \\ y_5 \end{bmatrix} = \begin{bmatrix} 0.0053 \\ 0.3359 \\ 0.8994 \\ 3152 \\ 872 \end{bmatrix}$$

By inverting the matrix the expression becomes:

$$\begin{bmatrix} y_1 \\ y_2 \\ y_3 \\ y_4 \\ y_5 \end{bmatrix} = \begin{bmatrix} 0.4618 & 0.3666 & 0.1262 & 0.0132 & -0.0299 \\ -0.3203 & -0.0512 & 0.0082 & 0.0053 & -0.0019 \\ -0.1044 & -0.4707 & -0.1103 & -0.0026 & 0.0262 \\ 0.1095 & 0.1062 & -0.2304 & -0.0107 & 0.0546 \\ 0.3224 & 0.6812 & 0.6477 & -0.0193 & 0.0835 \end{bmatrix} \begin{bmatrix} 0.0053 \\ 0.3359 \\ 0.8994 \\ 3152 \\ 872 \end{bmatrix}$$

Leading to the values:

$y_1 = 15.77$ mm Hence: $Q_1 = 170$ kN
$y_2 = 15.04$ $Q_2 = 282$
$y_3 = 14.39$ $Q_3 = 216$
$y_4 = 13.71$ $Q_4 = 154$
$y_5 = 12.79$ $Q_5 = 54$

The longitudinal bending moment diagram can now be obtained and is illustrated in fig. 3.14A.

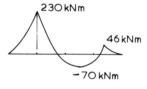
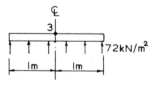

(A) Bending moment diagram for longitudinal direction

(B) Contact pressure at point 3 (for transverse moment)

Fig. 3.14 Foundation of varying width

As both the bending moment values and the foundation width vary, it is necessary to determine the required reinforcement at several points along the beam. For example, at nodal point 2 the total moment is 230 kNm, i.e. moment/metre width at point 2 is $230/2.5 = 92$ kNm/m.

Finite difference numerical techniques for foundations

For the determination of the required tranverse reinforcement it is necessary to calculate values for the transverse moments at selected points along the foundation.

The simplest way is to assume that the various values of q, which can be quickly obtained from the relationship $q = k_s y$, are distributed uniformly beneath the slab.

For example consider the transverse moment at section through point 3:

$$q_3 = 5 \times 14.4 = 72 \text{ kN/m}^2.$$

This pressure is assumed to act over a foundation width of 1.5 m (0.75 m on either side of point 3) see fig. 3.14B.

$$\text{Maximum transverse moment} = 72 \times 0.5 = 36 \text{ kNm}.$$

Obviously much greater accuracy is obtained if the analysis is carried out with the foundation divided into a greater number of sections.

Laterally loaded piles

Consider a vertical pile installed in a soil and at rest. The pressure acting on the periphery of the pile, p_p, will have a value approximately equal to the earth pressure at rest value if the pile was bored, or some higher value if the pile was driven.

Let us assume that it is necessary for the pile to experience a uniform displacement, y, throughout its length and to the right.

As soon as the pile begins to move to the right the pressure on the left hand face will drop to a very small value which can be assumed to be zero. On the right hand face of the pile, at the moment of movement, the pressure will immediately rise to a value p'_0 which will be greater than the earth pressure at rest value.

Hence, at the outset:

Lateral pressure on L.H.S. of pile ≈ 0

Lateral pressure on R.H.S. of pile, $p_p = p'_0$

After pile has experienced the displacement y:

Lateral pressure on L.H.S. of pile ≈ 0

Lateral pressure on R.H.S. of pile, $p_p = p'_0 + p$

where p = increase in pressure due to the displacement of the pile, y.

132 Elements of Foundation Design

It is generally found that the value of p is many times that of p_0'. It is therefore usually possible to neglect the term p_0' with little loss of accuracy.

Hence
$$p_p = p = k_h y$$

where k_h = modulus of horizontal subgrade reaction.

Determination of values for k_h

For stiff clays k_h is more or less independent of depth and the pressure resulting from a uniform displacement of the pile to the right is almost uniformly distributed down the right hand side of the pile.

However, as y increases with time (due to consolidation) so k_h tends to decrease and it is therefore necessary to use ultimate values of y when determining k_h.

In fact, for clays, the value taken for k_h in a pile design can be assumed to be equal to k_s, the modulus of vertical subgrade reaction for a horizontal beam resting on the same soil and having a width equal to the pile's width (or diameter). The determination of k_s has already been discussed.

With sands both y and k_h can be regarded as being independent of time in that deformation effects tend to occur almost instantaneously. However, as the modulus of elasticity of sand increases more or less proportionally with depth, so k_h increases proportionally with depth.

Let k_{h_1} = modulus of horizontal subgrade reaction obtained from a plate loading test on a square plate of dimensions 0.305×0.305 m.

Then
$$k_{h_1} \propto z$$

or
$$k_{h_1} = n_h z$$

where n_h = the constant of horizontal subgrade reaction for piles embedded in sand (units are MN/m^3).

Then
$$k_h = 0.305 \, n_h \frac{z}{B}$$

where k_h = the value of horizontal subgrade reaction at depth z, used in the design calculations of a pile of width (or diameter) B (in metres).

Typical values for n_h which can be used when no other information is available are set out below:

Typical values of n_h (MN/m³)			
	Consistency		
	Loose	Medium	Dense
Dry or moist sand	2.2	6.6	17.5
Submerged sand	1.25	4.4	10.5

Boundary conditions

The problem of a vertical pile subjected to lateral loading can be solved by finite differences in a similar manner to the case of a beam sitting on a Winkler foundation, provided that the boundary conditions are known or can be estimated.

Base of pile: the bottom of a pile can be displaced and can suffer rotation but, for all practical purposes, it can be assumed that no moment will develop.

Top of pile: The boundary condition at the top of a pile depends upon the form of fixity into the structure it is supporting. If there is a substantial pile cap then it can be assumed that there will be no rotation and that a moment will develop, although there can still be displacement. For a pile whose top is only lightly held it may be that no moment will develop.

It is generally simplest to ignore the boundary conditions at the top of a pile (as in the following examples).

Example 3.3

A concrete pile of square cross section (B = 446 mm) has been installed in a clay (k_h = 16.3 MN/m³) to a depth of 8 m. E_c = 20000 MN/m². A horizontal force of 15 kN will act at the top of the pile, which is 0.5 m above ground level. Determine the pile's displacement to depth profile and its bending moment diagram.

For the purpose of analysis divide the buried length of the pile into eight equal 1 m long sections (fig. 3.15B).

134 *Elements of Foundation Design*

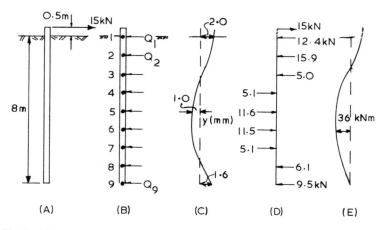

Fig. 3.15 Pile with lateral load in clay

$$EI = \frac{(0.446)^4}{12} \times 20\,000$$

$$= 65.9 \text{ MNm}^2$$

$Q_1 = k_h \times \text{area} \times y_1 = 16.3 \times 0.5 \times 0.75 \times y_1 = 6.1\,y_1$
$Q_2 = 16.3 \times 1.0 \times 0.75 \times y_2 = 12.2\,y_2 \ldots \text{etc.}$

Assume that all reactive forces $Q_1 \ldots Q_9$ are positive and act from right to left (fig. 3.15B).

Expressions for the moments at points 2 to 8 inclusive can now be found:

e.g. $M_6 = 15 \times 5.5 - 5Q_1 - 4Q_2 - 3Q_3 - 2Q_4 - Q_5$

and $-M_6 = \dfrac{EI}{a^2}(y_5 - 2y_6 + y_7)$

Equating these two expressions leads to:

$30.5\,y_1 + 48.8\,y_2 + 36.6\,y_3 + 24.4\,y_4 - 53.725\,y_5$
$\qquad\qquad\qquad\qquad + 131.85\,y_6 - 65.9\,y_7 = 82.5.$

Finite difference numerical techniques for foundations 135

This procedure gives seven of the required nine simultaneous equations. The remaining two equations are found by considering $\Sigma R = 0$ and $M_9 = 0$. The resulting set of equations, in matrix form, is:

$$\begin{bmatrix} 59.825 & -131.85 & 65.925 & 0 & 0 & 0 & 0 & 0 & 0 \\ 12.2 & -53.725 & 131.85 & -65.925 & 0 & 0 & 0 & 0 & 0 \\ 18.3 & 24.4 & -53.725 & 131.85 & -65.925 & 0 & 0 & 0 & 0 \\ 24.4 & 36.6 & 24.4 & -53.725 & 131.85 & -65.925 & 0 & 0 & 0 \\ 30.5 & 48.8 & 36.6 & 24.4 & -53.725 & 131.85 & -65.925 & 0 & 0 \\ 36.6 & 61.0 & 48.8 & 36.6 & 24.4 & -53.725 & 131.85 & -65.925 & 0 \\ 42.7 & 73.2 & 61.0 & 48.8 & 36.6 & 24.4 & -53.725 & 131.85 & -65.925 \\ 6.1 & 12.2 & 12.2 & 12.2 & 12.2 & 12.2 & 12.2 & 12.2 & 6.1 \\ 48.8 & 85.42 & 73.2 & 61.0 & 48.8 & 36.6 & 24.4 & 12.2 & 0 \end{bmatrix} \begin{bmatrix} y_1 \\ y_2 \\ y_3 \\ y_4 \\ y_5 \\ y_6 \\ y_7 \\ y_8 \\ y_9 \end{bmatrix} = \begin{bmatrix} -22.5 \\ 37.5 \\ 52.5 \\ 67.5 \\ 82.5 \\ 97.5 \\ 112.5 \\ 15.0 \\ 127.5 \end{bmatrix}$$

By inverting the matrix the values for $y_1 \ldots y_9$ can be found:

$y_1 = 2.04$ mm
$y_2 = 1.3$
$y_3 = 0.41$
$y_4 = -0.42$
$y_5 = -0.95$
$y_6 = -0.94$
$y_7 = -0.42$
$y_8 = 0.5$
$y_9 = 1.56$

Hence:
$Q_1 = 12.4$ kN
$Q_2 = 15.9$
$Q_3 = 5.0$
$Q_4 = -5.1$
$Q_5 = -11.6$
$Q_6 = -11.5$
$Q_7 = -5.1$
$Q_8 = 6.1$
$Q_9 = 9.5$

and
$M_1 = 7.5$ kNm
$M_2 = 10.1$
$M_3 = -3.2$
$M_4 = -21.5$
$M_5 = -34.7$
$M_6 = -36.3$
$M_7 = -27.0$
$M_8 = -11.4$
$M_9 = 0 \ (-2.5$ closure error$)$

(See fig. 3.15C, D and E)

136 Elements of Foundation Design
Example 3.4
A 3 m long concrete pile of square section (B = 290 mm) is driven into a dense sand. Ground water level occurs at a depth of 1.5 m below the surface of the sand. A clockwise moment of 50 kNm is to be applied to the top of the pile together with a 10 kN horizontal force which will act at a height of 0.75 m above the ground surface (fig. 3.16A). E_c = 22 1000 MN/m². Determine the pile's displacement to depth profile and its bending moment diagram.

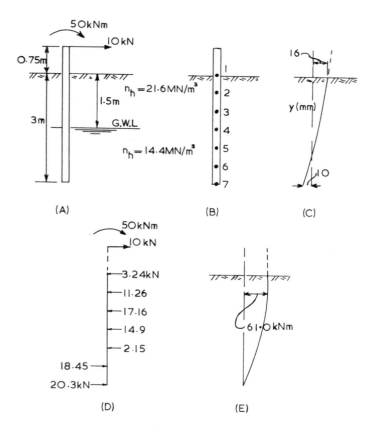

Fig. 3.16 Pile with lateral load in sand

Finite difference numerical techniques for foundations

For illustrative purposes the buried pile length has been divided into 6 equal parts 0.5 m long.

Values of k_h

Point 2. $k_h = 0.305 \times 21.6 \times 0.5 = 3.3 \text{ MN/m}^3$
Point 3. $k_h = 0.305 \times 21.6 \times 1.0 = 6.6 \text{ MN/m}^3$
Point 4. $k_h = 0.305 \times 21.6 \times 1.5 = 9.9 \text{ MN/m}^3$
Point 5. $k_h = 9.9 + 0.305 \times 14.4 \times 0.5 = 12.1 \text{ MN/m}^3$
Point 6. $k_h = 9.9 + 0.305 \times 14.4 \times 1.0 = 14.3 \text{ MN/m}^3$
Point 7. $k_h = 9.9 + 0.305 \times 14.4 \times 1.5 = 16.5 \text{ MN/m}^3$.

Assume that k_h at point 1, i.e. at the surface of the soil, equals the average value of k_h between the surface and point 2.

Then k_h for point $1 = \dfrac{0 + 3.3}{2} = 1.65 \text{ MN/m}^3$.

Then $Q_1 = k_h \times \text{area} \times y_1 = 1.65 \times 0.5 \times 0.25 y_1 = 0.206 y_1$
$Q_2 = k_h \times \text{area} \times y_2 = 3.3 \times 0.5 \times 0.5 y_2 = 0.825 y_2$
$Q_3 = 1.65 y_3$
$Q_4 = 2.475 y_4$
$Q_5 = 3.025 y_5$
$Q_6 = 3.575 y_6$
$Q_7 = 2.063 y_7$.

Expressions for the moments at points 2, 3, 4, 5 and 6 can be obtained:

For example, $M_5 = 50 + 10 \times 2.75 - 2Q_1 - 1.5Q_2 - Q_3 - 0.5Q_4$

and
$$M_5 = -\frac{EI}{a^2}(y_4 - 2y_5 + y_6).$$

$$EI = \frac{(0.29)^4}{12} \times 22\,100 = 13.02 \text{ MNm}^2$$

and
$$\frac{EI}{a^2} = \frac{13.02}{0.25} = 52.08.$$

This procedure will yield five equations and the remaining two are obtained from $M_7 = 0$ and $\Sigma R = 0$. The resulting matrix expression is set out below:

$$\begin{bmatrix} 51.977 & -104.16 & 52.08 & 0 & 0 & 0 & 0 \\ 0.206 & -51.668 & 104.16 & -52.08 & 0 & 0 & 0 \\ 0.309 & 0.825 & -51.255 & 104.16 & -52.08 & 0 & 0 \\ 0.412 & 1.238 & 1.65 & -50.842 & 104.16 & -52.08 & 0 \\ 0.515 & 1.65 & 2.475 & 2.475 & -50.567 & 104.16 & -52.08 \\ 0.618 & 2.063 & 3.3 & 3.713 & 3.025 & 1.788 & 0 \\ 0.206 & 0.825 & 1.65 & 2.475 & 3.025 & 3.575 & 2.063 \end{bmatrix} \begin{bmatrix} y_1 \\ y_2 \\ y_3 \\ y_4 \\ y_5 \\ y_6 \\ y_7 \end{bmatrix} = \begin{bmatrix} -62.5 \\ 67.5 \\ 72.5 \\ 77.5 \\ 82.5 \\ 87.5 \\ 10 \end{bmatrix}$$

On inverting the matrix the following values are obtained for y:

	Hence:		
$y_1 = 15.74$ mm	$Q_1 = 3.24$ kN	$M_1 = 57.6$ kNm
$y_2 = 13.65$	$Q_2 = 11.26$	$M_2 = 61.0$
$y_3 = 10.40$	$Q_3 = 17.16$	$M_3 = 58.7$
$y_4 = 6.02$	$Q_4 = 14.90$	$M_4 = 47.8$
$y_5 = 0.71$	$Q_5 = 2.15$	$M_5 = 29.5$
$y_6 = -5.16$	$Q_6 = -18.45$	$M_6 = 10.15$
$y_7 = -9.84$	$Q_7 = -20.3$	$M_7 = 0$

Check $\Sigma R = 0.04$. Check $M_1 = 50 + 10 \times 0.75 = 57.5$.

The various diagrams are shown in fig. 3.16.

References (chapter 3)

Boussinesq, J. (1885) *Application des Potentials à l'Étude de l'Équilibre de Mouvement des Solids Élastiques.* Gauthier-Villars, Paris.

Bowles, J. E. (1977) *Analytical and Computer Methods in Foundation Engineering.* Maidenhead: McGraw-Hill Inc.

Cheung, Y. K. (1977) 'Beams, slabs and pavements.' *Numerical Methods in Geotechnical Engineering,* chapter 5. Maidenhead: McGraw-Hill Inc.

Road Research Laboratory (1952) *Soil Mechanics for Road Engineers.* London: H.M.S.O.

Terzaghi, K. (1955) 'Evaluation of coefficients of subgrade reaction.' *Géotechnique* 5(4).

Terzaghi, K. and Peck, R. B. (1948) *Soil Mechanics in Engineering Practice.* Chichester: John Wiley and Sons Inc.

4. Reinforced Earth

Introduction

Reinforced earth is a relatively new civil engineering material which has only been used commercially for the past twelve years or so. Its main use has been in the construction of earth retaining structures and bridge abutments but it is now being adopted into the field of foundation stabilisation and its possible future use might even include the strengthening of cuttings.

This chapter is restricted to the use of reinforced earth in earth retaining structures a simple form of which is illustrated in fig. 4.1. Brief descriptions of the components listed in the figure are set out below.

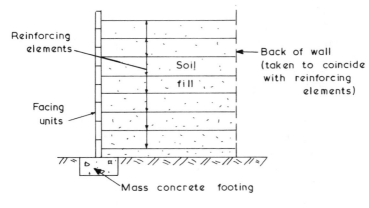

Fig. 4.1 Typical reinforced earth retaining wall

Soil fill

Three types of soil fill are available:

Frictional fill

Most practical reinforced earth structures that have been built have employed frictional, or granular, soil. The reasons are not far to seek as a granular soil not only has good frictional resistance but is also free draining and generally less corrosive than a cohesive soil.

Elements of Foundation Design

Cohesive frictional soil

The Department of Transport Memorandum (1978) permits the use of silty soils as reinforced earth fill, subject to certain conditions listed later.

Cohesive soil

Research is being carried out into the possibility of using clay as a fill for reinforced earth (Murray, 1977; Boden *et al.*, 1977). As clay is probably the most common soil encountered in the United Kingdom, encouraging results from such research would be of interest but, as Boden points out, any potential benefits arising from the local availability of such soil could be outweighed by the penalties that might arise, such as difficulty in handling, development of pore water pressures and the greater risk of corrosion.

Reinforcing elements

Almost all the reinforced earth structures that have been built so far have employed metallic reinforcing elements, the most common being galvanised steel. Each element is a thin strip of metal, typically 50–100 mm wide and up to 9 mm thick, but of several metres in length.

Other metals from which reinforcement strips have been prepared include stainless steel, aluminium, aluminium alloy and copper.

Reinforcing elements made from reinforced concrete in the form of thin prestressed planks have been used occasionally.

Plastic materials show promise for the future, although more research is necessary. Two main types are presently available:

Fibre reinforced plastic–consists of glass filaments embedded in polyester resin.

Paraweb–polyester filaments embedded in polyethylene.

Facing units

At a free boundary of a reinforced structure it is necessary to provide some form of barrier so that the soil is contained. This skin can be either flexible or stiff but it must be strong enough to hold back the local soil and to allow fastenings for the reinforcement to be attached.

The facing of a reinforced earth structure is usually prefabricated from units which are small and light enough to be manhandled for quick and easy construction. The units are generally made from steel, aluminium, reinforced concrete or plastic.

The facing units require a small foundation from which they can be built, generally consisting of a trench filled with mass concrete, giving a footing similar to those used in domestic housing.

Principle of reinforced earth

Consider a semi-infinite mass of cohesionless soil at rest. If the surface of the soil is horizontal then, at depth h below the surface:

$$\text{Vertical stress} = \gamma h$$
$$\text{Lateral stress} = K_0 \gamma h$$

where K_0 = coefficient of earth pressure at rest; γ = unit weight of the soil.

According to Jaky (1944), for both normally consolidated clays and compacted soils, $K_0 \approx 1 - \sin \phi$ where ϕ = the angle of friction of the soil.

If the soil is allowed to expand laterally, the horizontal stress, $K_0 \gamma h$, reduces to a limiting (or failure) value, $K_a \gamma h$,

where K_a = coefficient of active earth pressure

$$= \frac{1 - \sin \phi}{1 + \sin \phi} = \tan^2\left(45° - \frac{\phi}{2}\right).$$

If the soil is compressed laterally, the horizontal stress increases to limiting value, $K_p \gamma h$,

where K_p = coeficient of passive earth pressure

$$= \frac{1 + \sin \phi}{1 - \sin \phi} = \tan^2\left(45° + \frac{\phi}{2}\right).$$

The three stress states are illustrated in fig. 4.2.

Now reconsider the soil mass with horizontal reinforcement strips embedded within it (fig. 4.3).

Consider a soil layer between two adjacent reinforcement strips. If enough friction is developed, the top and bottom of the layer will be attached to the reinforcements. If the strips are close enough then the whole soil layer will be more or less constrained and the maximum strain that it can experience in the direction of the reinforcements will be of the order of the strain in the reinforcements.

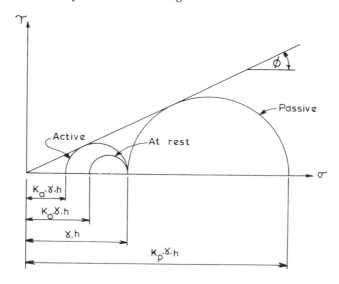

Fig. 4.2 Variation of lateral pressure

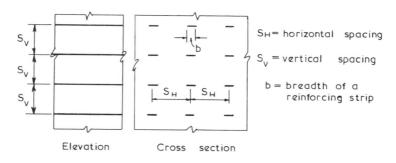

Fig. 4.3 Arrangement of reinforcement strips

The types of reinforcement material available are discussed later but all have a Young's modulus much greater than that for the soil so that the resulting strains in the soil will be so small that the soil is essentially at rest and the lateral pressure within it can be assumed to be equal to $K_0 \gamma h$.

This is the fundamental idea of reinforced earth, developed by Vidal in 1966. Reinforced earth, therefore, is a combination of soil, which is weak in tension, and reinforcing elements which can carry the tensile forces transmitted from the soil. The composite material is strong in compression and has tensile strength in the direction of the reinforcements.

In this respect it is somewhat analogous to reinforced concrete.

It must be remembered that the tensile strength in reinforced earth is directional. If the soil is compressed laterally (instead of vertically) the horizontal reinforcement would have no effect. For such a case, in order to obtain tensile strength in the vertical direction, it would be necessary to insert vertical reinforcement.

Fundamental theory

A comprehensive treatment of the relevant theory was set out by Schlosser and Vidal (1969). A synopsis of the main points is set out below.

Principal stresses and strains within the soil

An isotropic and homogeneous granular soil mass, without reinforcement, has identical properties in all directions. The strength envelope of such a soil consists of two lines, passing through the origin, inclined at angles of $\pm\phi$ to the normal stress axis (fig. 4.4A).

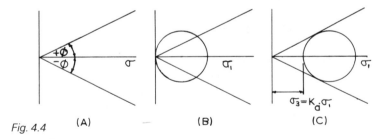

Fig. 4.4 (A) (B) (C)

If a cube of such a soil is subjected to unconfined compression, it will fail as the stress circle cuts through the strength envelope (fig. 4.4B). For the cube to be stable a lateral stress, σ_3, not less than $K_a\sigma_1$, must be applied (fig. 4.4C).

If the soil is now reinforced in the direction of σ_3, the lateral strain in the soil, ϵ_3, is very small – the composite material is laterally confined in the direction of σ_3 and the value of this stress will be about $K_0\sigma_1$ (fig. 4.5).

146   *Elements of Foundation Design*

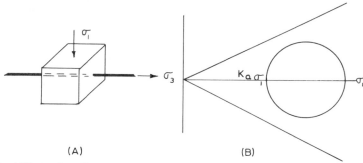

Fig. 4.5 Lateral confinement

Theoretically, soil failure can never occur provided it is laterally confined (fig. 4.5B) as the stress circle is always within the strength envelope, no matter what the value of σ_1. Failure can only occur if the reinforcement breaks or pulls out of the soil.

If we assume that the cube faces are of unit area then lateral stress = lateral force = $K_0 \sigma_1$.

This is the force that is transferred from the soil into the reinforcement.

Hence the tensile stress in the reinforcement = $\dfrac{K_0 \sigma_1}{A_r}$

where A_r = cross-sectional area of reinforcement.

∴ strain in reinforcement = $\dfrac{K_0 \sigma_1}{A_r E_r}$

where E_r = Young's modulus for reinforcement material.

Hence, for the elemental cube shown in fig. 4.5A the resulting strains are:
 (i) A vertical strain, $\epsilon_1 = (\sigma_1/E_s)$ where E_s = Young's modulus for the soil.
 (ii) A lateral strain, ϵ_3 (in the direction of the reinforcement) of a small magnitude and equal to $(K_0 \sigma_1/A_r E_r)$.
 (iii) A lateral strain, ϵ_2–this will also be of small magnitude as a wall of any reasonable length operates in a state of plane strain.

A semi-infinite mass of soil with a horizontal surface can be considered as being at rest, whether or not it is reinforced with horizontal reinforcement. As lateral deformations are zero, the lateral pressures are equal to $K_0\sigma_1$ where σ_1, the major principal stress, is equal to the vertical stress γh.

If a vertical cut is made through such a soil there will be a considerable change in the lateral stress if the soil is not reinforced. To prevent a change in stress it would be necessary to apply a hydrostatic pressure distribution, p, along the cut such that $p = K_0\gamma h$ (fig. 4.6).

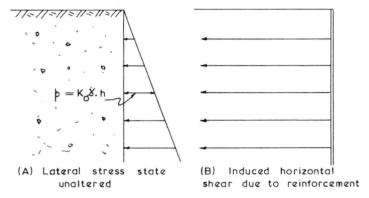

(A) Lateral stress state unaltered

(B) Induced horizontal shear due to reinforcement

Fig. 4.6

If the soil mass had been reinforced by horizontal strips then the hydrostatic pressure would have been effectively applied to the vertical surface of the cut as each reinforcement strip would have a tensile force induced into it. However, the only way for tension to be induced into the strips is for them to be bonded to the soil and this can only happen if horizontal shear stresses develop along the surfaces of the strips and hence within the soil. It is seen therefore that the soil mass is subjected to a system of horizontal shear stresses. Hence, with the reinforcement, the lateral strains are vertically unaltered but there the soil mass is no longer in a state of rest, $\sigma_3 \neq K_0\sigma_1$, $\sigma \neq \gamma h$. The stress state is illustrated in fig. 4.7 which shows that, if the shear stresses are large enough, a state of failure can be approached.

Rankine and Coulomb theories of active earth pressure

Full descriptions of these theories are available in any soils text book. Briefly, the Rankine theory considers the equilibrium of an element in the soil mass and deduces that a network of potential

148 *Elements of Foundation Design*

failure planes, at $(\phi/2) + 45°$ to the horizontal, exists behind the wall (fig. 4.8A), for a wall supporting dry sand with a horizontal surface.

The Coulomb assumption considers the whole of the soil mass retained and assumes that failure will occur by a wedge of soil sliding down a failure surface. For practical purposes the failure surface can be considered as a straight line fill and, for a sand behind a wall, is inclined at $(\phi/2) + 45°$ to the horizontal (fig. 4.8B).

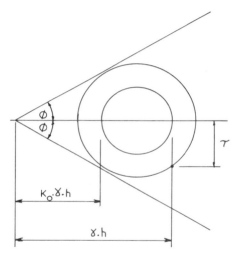

Fig. 4.7 Effect of shear stresses

The active pressure distribution behind a retaining wall approximates to either the Rankine or the Coulomb condition, depending upon the amount and type of yield that the wall has experienced (Smith 1978).

A wall can yield in one of two ways: either by rotation about its lower edge or by sliding forward. Provided it yields sufficiently, a state of active pressure is reached and the total thrust on the back of the wall is P_a. The distribution that gives this P_a value can be very different and depends upon the way in which the wall has yielded.

Consider first a wall that is unable to yield (fig. 4.9A). The soil is at rest and the pressure distribution is represented by line AC.

Consider now that the wall fails by rotation about its lower edge until the total active thrust is P_a (fig. 4.9B). This results in conditions

Reinforced earth 149

approximating to the Rankine theory and is known as the totally active case.

Suppose, however, that the wall yields by sliding forward until active thrust conditions are achieved and the total thrust again equals P_a. A forward displacement hardly disturbs the upper layers of soil so that the top of the pressure diagram is similar to the earth pressure at rest diagram. As the total thrust on the wall is the same as for rotational yield, the pressure distribution must be similar to the line AE in fig. 4.9C. These conditions correspond to the Coulomb theory.

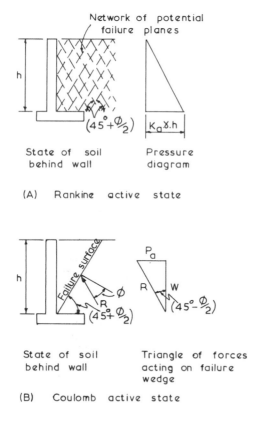

Fig. 4.8 Rankine and Coulomb active states

150 *Elements of Foundation Design*

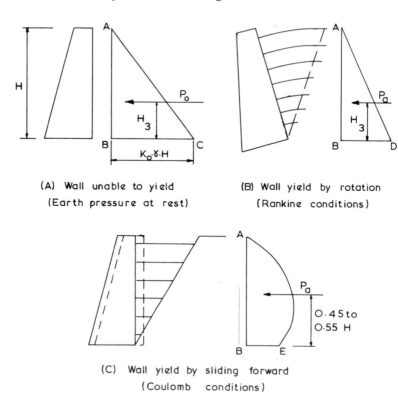

(A) Wall unable to yield (Earth pressure at rest)

(B) Wall yield by rotation (Rankine conditions)

(C) Wall yield by sliding forward (Coulomb conditions)

Fig. 4.9 Influence of wall yield on lateral pressure distributions behind a retaining wall

If we consider a frictional fill, angle of friction ϕ, unit weight γ, and assume a frictionless wall of height, H, then both theories provide the same value for the total active thrust, P_a, provided that the surface of the fill is horizontal and parallel with the back of the wall.

Rankine. $K_a = \tan^2 \left(45° - \dfrac{\phi}{2} \right)$

P_a = area of pressure distribution diagram

$= \tfrac{1}{2} K_a \gamma H^2 = \tfrac{1}{2} \gamma H^2 \tan^2 \left(45° - \dfrac{\phi}{2} \right)$

Coulomb. Area of sliding triangular wedge $= \frac{1}{2}H^2 \tan\left(45° - \frac{\phi}{2}\right)$

Weight of sliding wedge, W $= \frac{1}{2}\gamma H^2 \tan\left(45° - \frac{\phi}{2}\right)$

From triangle of forces: $P_a = W \tan\left(45° - \frac{\phi}{2}\right)$

$\qquad\qquad\qquad\qquad\quad = \frac{1}{2}\gamma H^2 \tan\left(45° - \frac{\phi}{2}\right)$

Prediction of forces in reinforcement strips

Whether or not the theories of Rankine and Coulomb should be used in an unmodified form in the analysis of reinforced earth has been argued amongst researchers for several years.

Some researchers consider that to apply the classic theories of earth pressure to reinforced earth is quite erroneous as the reinforcement inclusions completely invalidate the notion of isotropic homogeneity.

They maintain, not unreasonably, that as reinforced earth is an entirely new and different material there is a need for a complete understanding of its behaviour so that realistic design formulae can eventually be evolved.

However, at this stage of development, there is little choice for designers but to use the Rankine and Coulomb assumptions. It should be remembered that, literally thousands of reinforced earth structures have been designed with these theories, have been constructed and have proved satisfactory.

Let us consider the reinforced earth wall of fig. 4.1 and deduce expressions for the maximum tensile forces in the reinforcing elements. Both theories will be used, in their most simple form, but it should be noted that there are other approaches, a summary of which has been prepared by Symons (1973).

Coulomb theory

The theory assumes that the structure will fail by the strips either snapping or being pulled out and causing the wedge of reinforced fill, ABC, to slip along some plane AC inclined at θ to the horizontal (fig. 4.10A).

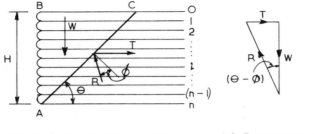

(A) Coulomb failure wedge (B) Triangle of forces on wedge

Fig. 4.10 Coulomb theory

The various forces acting on the wedge are:

W = weight of wedge = $\frac{1}{2}\gamma H^2 \tan(90° - \theta)$
R = reaction on plane of failure (at ϕ to normal)
T = total tensile force in strips/m length of wall.

From force diagram T = W $\tan(\theta - \phi)$ (fig. 4.10B)
= $\frac{1}{2}\gamma H^2 \tan(90° - \theta) \tan(\theta - \phi)$.

For maximum value of T, $\theta = (45° + \phi/2)$

i.e.

$T_{max} = \frac{1}{2}\gamma H^2 \tan(45° - \phi/2) \tan(45° - \phi/2) = \frac{1}{2}\gamma H^2 \tan^2(45° - \phi/2)$

i.e. $T_{max} = \frac{1}{2}K_a \gamma H^2$/m length of wall.

This is the resultant of all the forces in the strips and must be distributed amongst them. This is achieved by assuming a triangular distribution, the forces increasing with depth (fig. 4.11).

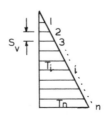

Fig. 4.11

Reinforced earth

Let n = total number of strips and consider strip i, counting down from top.

$$\frac{T_i}{iS_v} = \frac{T_n}{nS_v} \qquad \therefore T_i = \frac{T_n i}{n}$$

Now $\quad T_1 + T_2 + T_3 + \ldots T_{(n-1)} + T_n = T$

$$\therefore \frac{T_n}{n} + \frac{2T_n}{n} + \frac{3T_n}{n} + \ldots \frac{(n-1)T_n}{n} + \frac{nT_n}{n} = T$$

$$\therefore T_n (1 + 2 + 3 + \ldots (n-1) + n) = nT.$$

Now $\quad \Sigma n = \dfrac{n(n+1)}{2}$

$$\therefore T_n = \frac{2nT}{n(n+1)} = \frac{2T}{n+1}$$

$$\therefore T_i = \frac{2T_i}{n(n+1)} = \frac{i}{n(n+1)} K_a \gamma H^2. \tag{1}$$

Rankine theory

Consider a typical reinforcing element at level i (fig. 4.12). It can be assumed that the soil associated with the element will consist of half the upper layer of soil and half the lower layer, i.e. a strip of soil, of thickness S_v, extending from level $(i - \tfrac{1}{2})$ to $(i + \tfrac{1}{2})$, as shown in fig. 4.12.

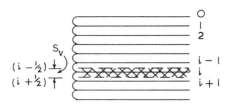

Fig. 4.12

Vertical pressure acting on soil at level $(i - \tfrac{1}{2}) = \gamma S_v (i - \tfrac{1}{2})/m$ length of wall.

Assuming that this stress is a principal stress, the active pressure at this level, $p_{a(i - \frac{1}{2})} = K_a \gamma S_v (i - \tfrac{1}{2})$.

The vertical pressure at level $(i + \tfrac{1}{2}) = \gamma S_v (i + \tfrac{1}{2})/m$ length of wall.

Hence the active pressure at this level, $p_a(i + \frac{1}{2}) = K_a \gamma S_v (i + \frac{1}{2})$.

The pressure diagram for this strip of soil is therefore as shown in fig. 4.13.

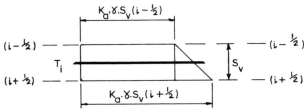

Fig. 4.13 Active pressure diagram for strip i

Tension in reinforcing element, T_i = area of active pressure diagram

$$\therefore T_i = S_v K_a \gamma S_v \left[\frac{(i - \frac{1}{2}) + (i + \frac{1}{2})}{2} \right]$$

$$= i K_a \gamma S_v^2 \qquad (2)$$

Examination of equation (1) for T_i, shows that, for large values of n, it approximates to:

$$T_i = i K_a \gamma \left(\frac{H}{n} \right)^2$$

But
$$\frac{H}{n} = S_v$$

so that, for large values of n, equation (1) tends towards equation (2).

Equation (2) can therefore be used as a first approximation in order to obtain a value for the tensile force/m length of wall at the level of strip i. It is often best written in the form:

$$T_i = K_a S_v \sigma_v$$

where $\sigma_v = \gamma i S_v$ = vertical stress at level i.

The maximum tension line

It has been found, from both model tests and measurements on constructed works, that the tensile force in a reinforcement strip varies. It generally has a low (even a zero) value at the facing unit, reaches a maximum value a short distance from the facing and then tends towards zero at the unattached end (fig. 4.14A).

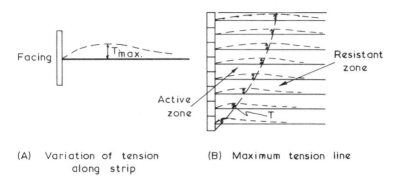

(A) Variation of tension along strip

(B) Maximum tension line

Fig. 4.14 Maximum tension line

If the points of maximum tensile force in each strip are joined then the imaginary line formed is known as the maximum tension line (fig. 4.14B). It extends in a curve from the facing at the base of the wall to cut the surface of the fill at some distance back from the facing. It is assumed that this line divides the reinforced fill into two zones:

> An active zone in which the shearing stresses from the soil, on to the strips, act towards the facing, i.e. there is a tendency for the reinforcement to be pulled forward towards the facing.

> A resistant zone in which the shear stresses act away from the facing and tend to hold the reinforcement in the soil.

T_i, as obtained from equation (2) is assumed to be the maximum tensile force in the strip and therefore occurs at some distance back from the facing.

Failure of a reinforcing element
The tensile force in a reinforcing strip will tend to cause failure in one of two ways:

> Tensile failure—snapping of the strip

> Bond failure—the slipping of the strip within its surrounding soil.

Tensile failure

The ultimate resistance of a reinforcing element to an axial tensile stress is equal to the ultimate axial tensile stress that the material can withstand, p_{ult}, times the cross sectional area of the strip.

$$R_t = p_{ult} \times b \times t$$

where b = width of reinforcing element
t = thickness of reinforcing element.

Dividing p_{ult} by a suitable factor of safety gives the permissible axial tensile stress, p_{at}.

Hence, when considering tensile failure effects: $T_i \not> p_{at}$ bt.

Typical p_{at} values are:

	N/mm²
Aluminium	120
Galvanised mild steel	120–190
Copper alloy	170
Stainless steel	120–220

Bond failure

Just as a reinforcing bar in concrete requires a minimum bond length, so a reinforcing element in reinforced earth must be long enough to prevent slipping and pull out.

For a reinforcing element the bond resistance between it and the soil will be provided by one of three means, depending upon the type of soil fill:

Frictional fill–by frictional resistance.

Cohesive fill–by adhesion.

Cohesive frictional fill–by both frictional resistance and adhesion.

Frictional resistance

Consider a strip at depth h_i and assume that the coefficient of friction between the soil and the reinforcing element is μ.

Normal stress acting on strip = vertical stress = γh_i.
∴ Normal force acting on strip = normal stress × area = γh_i bL.
∴ Total frictional resistance available from strip = $2\gamma h_i$ bLμ.

Note. Just as the peripheral area of a reinforcing bar in concrete is considered to provide bond resistance so the total area, i.e. the upper and lower surfaces, of the strip is considered to provide bond resistance.

Adhesion

Total resistance available from adhesion in a cohesive soil $= 2c_r bL$
where $c_r =$ unit adhesion available along the length of the reinforcing element.

Note. μ and c_r values are obtained from shear box tests (described later) but μ is, typically, approximately equal to $\frac{1}{2} \tan \phi$ and c_r will lie somewhere between $\frac{1}{2} c_u$ and $1.0 c_u$.

Friction and adhesion
Generally, for a $c - \phi$ fill:
Bond resistance of a reinforcing element $= 2(\gamma h_i \mu + c_r)bL$.

Design criteria

Two major aspects must be considered when designing a reinforced earth retaining wall:

Internal Stability	(a) Tensile resistance
	(b) Bond resistance
	(c) Pressure on facing.
External stability	(a) Rotational type failure
	(b) Failure by sliding as a rigid body
	(c) Bearing failure of supporting soil.

Internal stability

There are two major approaches by which internal stability can be checked:
  (i) Analysis involving the local stability of single reinforcing elements
  (ii) Analysis involving the overall stability of blocks of the soil comprising the wall.

Note. Neither assumption approaches the actual operating conditions within a typical structure and opinion appears to be equally divided as to which is the better approach.

Method (i) has the advantage of being quick to use and general practice appears to be to design a reinforced earth retaining wall by

158 *Elements of Foundation Design*

that method and then to check the result by method (ii), modifying the initial design if necessary.

Method (i) Check on local stability

For a wall such as illustrated in fig. 4.1, with a granular fill, the maximum tensile force in the reinforcing elements at level i is obtained from equation (2):

$T_i = iK_a \gamma S_v^2/m$ run of wall (minus $2cS_v\sqrt{K_a}$ if cohesive frictional fill).

Treatment of uniform surcharge on top of wall.

If the wall is uniformly loaded by a surcharge of w_s/m^2 there will be an increase in T_i due to the uniform pressure distribution $K_a w_s$ induced within the soil.

Then $T_i = K_a S_v (\gamma i S_v + w_s)/m$ length of wall (minus $2cS_v\sqrt{K_a}$ if cohesive frictional fill).

Treatment of line load on top of wall.

Schlosser and Long (1974) carried out measurements in reinforced earth fill materials, both on models and on an experimental wall, to gauge the effect of a line load acting on the surface of the fill and found that the load spread through the reinforced earth at a slope between 2/1 and 1/1. They proposed the following design assumptions:

If line load = S_L and its point of application is distance 'd' back from the facing, then (referring to fig. 4.15):

Vertical stress at level i due to $S_L = \dfrac{S_L}{d + \frac{1}{2}h_i}$.

Hence, increase in T_i due to $S_L = K_a S_v \dfrac{S_L}{d + \frac{1}{2}h_i}$.

Note. Although the dimension 'd' can be of value, for the purpose of load spreading only, it cannot be taken as greater than $\frac{1}{2}h_i$ in the above formula.

The method becomes conservative when, as is usually the case, the line load can be considered as being applied through a continuous footing or the equivalent, e.g. railway track.

If the load is applied concentrically then the spread length at level i, D_i, will be $(d + B + \tfrac{1}{2}h_i)$ where B = width of footing (fig. 4.15B).

If S_L is applied eccentrically to the footing, it can be assumed that the bearing pressure distribution beneath it is trapezoidal. Then, for simplicity, one can assume that the maximum value of bearing pressure applies uniformly beneath the footing, i.e. that a uniform vertical pressure of

$$\frac{S_L}{B}\left(1 + \frac{6e}{B}\right)$$

acts at the top of the wall over the distance 'B' (fig. 4.15C).

At depth h_i, the spread length, D_i, is again $(d + B + \tfrac{1}{2}h_i)$ and the vertical pressure, σ_{vi}, can be taken as $[p_{max} \times (B/D_i)]$

or
$$\sigma_{vi} = \frac{S_L}{D_i}\left(1 + \frac{6e}{B}\right).$$

Hence, increase in T_i due to S_L

$$= K_a S_v \frac{S_L}{D_i}\left(1 + \frac{6e}{B}\right)$$

where d in the expression for D_i has the same limitation as noted above.

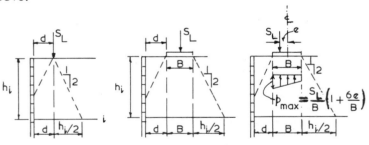

(A) Unspread concentrated load (B) Concentric load (C) Eccentric load

Fig. 4.15 Treatment of line load

Treatment of horizontal force at top of wall

Quite often the traction forces of machinery running along the top of the wall can induce a horizontal shear force, F_L, which is applied through some form of foundation of width B, and may be regarded as continuous along the length of the wall (fig. 4.16).

160 *Elements of Foundation Design*

Treatment is to assume that the force F_L is carried by the reinforcing elements which are contained in a Coulomb wedge passing through the edge of the foundation.

It is assumed that there is a triangular distribution of forces, decreasing with depth.

Let h = height of Coulomb wedge.

Then
$$h = \frac{d + B}{\tan(45° - \tfrac{1}{2}\phi)}.$$

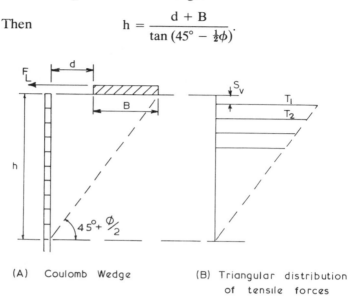

(A) Coulomb Wedge

(B) Triangular distribution of tensile forces

Fig. 4.16 Treatment of horizontal force

As tensile force distribution is assumed to be triangular (fig. 4.16B):

$$\frac{T_1}{h - S_v} = \frac{T_i}{h - iS_v}$$

$$\therefore T_i = \frac{T_1(h - iS_v)}{(h - S_v)}. \tag{A}$$

Now $\quad T_1 + T_2 + T_3 + \ldots T_n = F_L$

$\therefore T_1[(h - S_v) + (h - 2S_v) + (h - 3S_v) + \ldots (h - nS_v)] = F_L(h - S_v)$

or $\quad T_1 \left[nh - \dfrac{S_v}{2} n(n + 1) \right] = F_L(h - S_v).$

Now
$$n = \frac{h}{S_v}.$$

When this substitution is made the expression for T_1 becomes:

$$T_1 = \frac{2S_v F_L}{h}$$

and, substituting in equation (A) for T_i:

$$T_i = \frac{2S_v F_L}{h}\left(\frac{h - iS_v}{h - S_v}\right) = \frac{2S_v F_L}{h}\left(\frac{h - h_i}{h - S_v}\right).$$

The $- S_v$ term in the denominator can be neglected with little loss in accuracy:

Hence
$$T_i = \frac{2S_v F_L}{h}\left(1 - \frac{h_i}{h}\right).$$

Method (ii) Block stability analysis
As has been discussed, the maximum tension line more or less marks the division of a reinforced earth fill into an active zone and a resistant zone.

The block stability analysis assumes that the active zone consists of a failure wedge tending to pull away from the rest of the fill whereas the resistant zone, by gripping the ends of the reinforcing elements in the fill is 'anchoring' the active zone in position.

Form of the failure wedge (or active zone)
As a result of extensive observations by many research workers on both models and actual structures it is now generally accepted that, just before failure, a reinforced earth retaining wall develops some form of failure wedge and that the failure surface tends to follow the maximum tension line.

Obviously, if the exact shape of the failure wedge were known it would lead to a relatively straightforward design method. However, as Schlosser points out, the boundary between the active and resistant zones is variable and depends upon the geometry, the stress values, settlements within the subgrade and, possibly the most significant, the value of the factor of safety for the particular structure.

It must be remembered that the maximum tension line in a reinforced earth structure does not have a unique position: it varies, daily, with the loading and the state of the weather.

162 Elements of Foundation Design

For the almost theoretical case of a wall similar to that in fig. 4.1, i.e. with an unloaded horizontal surface, the maximum tension line is nearly vertical for about one half the height of the wall (fig. 4.17B), an idealised form of which, suitable for design purposes, appears in fig. 4.17C.

As can be seen, the volume of soil in the active zone of such a wall is considerably less than that contained in the Coulomb wedge (fig. 4.17A). Obviously, if the reinforcing elements are all of the same length, a much larger aggregate length is available for bond if the active zone is as fig. 4.17C than if the active zone approximates to the Coulomb wedge of fig. 4.17A. The question is whether it is safe to use the assumption of fig. 4.17C for all the loadings that can be applied to a reinforced earth retaining wall.

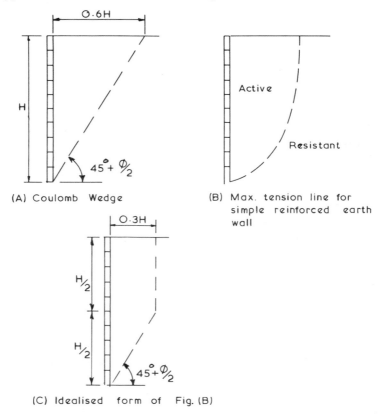

(A) Coulomb Wedge

(B) Max. tension line for simple reinforced earth wall

(C) Idealised form of Fig. (B)

Fig. 4.17 Form of failure wedge

In view of Schlosser's findings such an assumption could be foolhardy. Murray (1977) observes that for the more complex situations involving concentrated loads and sloping backfill the plane surface of failure is unlikely; it is more likely to be curved. Murray advocates, in view of the uncertainties involved, that the most direct approach to a solution would involve the analysis of a number of trial surfaces in order to determine the maximum thrust, as in normal retaining wall design. This approach has been adopted in example 4.1 below.

Although the trial failure surfaces can be of any shape the use of straight lines makes the calculations easier. The best way to describe the technique is to work through a typical example of a reinforced earth retaining wall design.

Example 4.1
A proposed reinforced earth retaining wall is shown in fig. 4.18. The wall is to be 10.5 m high and the 200 kN line load can be assumed to be applied centrally on its 1 m wide base. The reinforcing elements are to be of galvanised mild steel (permissible tensile stress = 125 N/mm^2) and the properties of the soil fill can be assumed to be $\phi = 30°$; $\gamma = 20$ kN/m^3. Maximum allowable bearing pressure = 300 kN/m^2. Design the wall.

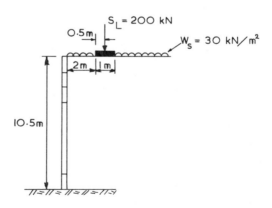

Fig. 4.18 Reinforced earth design example

Preliminary estimate of reinforcement

The total number of elements and their cross section must first be roughly estimated. This can be done by use of the Rankine theory. Ignoring the effects of S_L for the moment:

Total horizontal thrust within the wall $= T = \frac{1}{2} K_a \gamma H_e^2$

where H_e = equivalent height of wall allowing for the surcharge w_s.

$$H_e = 10.5 + \frac{30}{20} = 12.0 \text{ m}$$

$$K_a = \frac{1 - \sin 30°}{1 + \sin 30°} = 0.333$$

$$T = \frac{0.333 \times 20 \times 12^2}{2} = 480 \text{ kN/m length of wall.}$$

Try reinforcing elements of cross section 80 mm wide by 3 mm thick.

Then tensile resistance of an element,

$$R_t = \frac{3 \times 80 \times 125}{1000} = 30 \text{ kN.}$$

∴ Total number of strips required/m length of wall $= \dfrac{480}{30} = 16$.

Very little guidance is available as to the maximum dimensions that can be chosen for S_v and S_H. The maximum value of S_v that appears to have been used in actual structures is 1.0 m. It is suggested that 0.75 m should be regarded as the maximum values for both S_v and S_H.

At this early stage, bearing in mind that S_L has so far been ignored, put $S_v = 0.75$ m. Then number of rows of reinforcing elements $= 10.5/0.75 = 14$.

Depending upon the type of facing units and their fastenings the final arrangement of reinforcement will appear in cross section either as fig. 4.19A or as fig. 4.19B.

The design calculations that follow are for fig. 4.19B (which is more conservative and the results of which can be applied to the arrangement shown in fig. 4.19A).

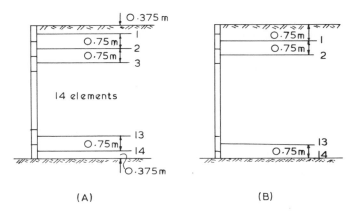

Fig. 4.19 Possible reinforcement arrangements for design example

Length of reinforcing strips

It has generally been established that the minimum length of reinforcing elements, to satisfy bond requirements, is approximately 0.8 times the total height of the wall. The actual length of the reinforcement is, of course, related to the type of soil and the form of loading but, initially let $L = 0.8\,H$.

i.e. length of reinforcing elements $\not< 0.8 \times 10.5 = 8.5$ m, say.

It is now possible to check on these initial choices and to design the wall.

Internal stability

(i) Local stability

This is best done in tabular form (table 4.1):

The maximum tensile force (66.1 kN/m run of wall) occurs at the level of strip 14.

The tensile resistance of a single strip, $R_t = 30.0$ kN. Hence, if the horizontal spacing, S_H, is maintained at 0.75 m, the tensile resistance of the wall would be $\frac{4}{3} \times 30 = 40$ kN/m run of wall.

At this stage one can vary any of the variables, i.e. S_H, S_v or the cross section of the reinforcing element.

Elements of Foundation Design

Table 4.1

Element	$\sigma_v = \gamma i S_v$ (kN/m²)	$\sigma_v + w_s$ (kN/m²)	D_i (m)	$\Delta\sigma_v S_L =$ S_L/D_i (kN/m²)	σ_{v_t} (kN/m)	$K_a S_v \sigma_{v_t}$ (kN/m)	Bond resist. (kN/m)
1	15	45	1.75	114.3	159.3	39.8	22.2
2	30	60	2.25	88.9	148.9	37.2	29.6
3	45	75	3.25	61.5	136.5	34.1	37.0
4	60	90	4.0	50.0	140.0	35.0	44.4
5	75	105	4.75	42.1	147.1	36.8	51.8
6	90	120	5.25	38.1	158.1	39.5	59.2
7	105	135	5.63	35.5	170.5	42.6	66.6
8	120	150	6.0	33.3	183.3	45.8	74.0
9	135	165	6.38	31.4	196.4	49.1	75.0
10	150	180	6.75	29.6	209.6	52.4	75.0
11	165	195	7.13	28.1	223.1	55.8	75.0
12	180	210	7.50	26.7	236.7	59.2	75.0
13	195	225	7.88	25.4	250.4	62.5	75.0
14	210	240	8.25	24.2	264.2	66.1	75.0

If S_v is retained at 0.75 m and if the reinforcement is increased to 100 mm wide,

$$R_t = \frac{100 \times 3 \times 125}{1000} = 37.5 \text{ kN.}$$

Then, if S_H is reduced to 0.5 m,

the tensile resistance available/metre run of wall = $2 \times 37.5 = 75$ kN

which is greater than 66.1 kN.

Hence, at this stage, a suitable arrangement of reinforcement would be:

$$S_v = 0.75 \text{ m}; \quad S_H = 0.5 \text{ mm};$$

dimensions of strip = 100 mm wide by 3 mm thick.

Bond resistance

The bond resistance available from a particular reinforcement strip can be obtained from the formula:

$$\frac{2b\mu\sigma_v L}{2} = b\mu\sigma_v L$$

Note. The 2 in the denominator is an overall factor of safety that is applied to ensure that adequate bond exists at working loads.

σ_v varies for each reinforcing element so the bond resistance available at a particular level is best presented in tabular form (see table 4.1). The values were calculated using the above formula with L = 8.5 m, b = 100 mm and $\mu = 0.5 \tan \phi = 0.29$. The value used for the vertical stress is that due to the weight of the soil and the uniform surcharge on the top of the wall ($\sigma_v + w_s$) in the table.

Note. To be logical, it is not possible to have an element with a bond resistance greater than its tensile resistance. Whenever the calculated bond resistance of an element exceeds 37.5 kN (the R_t value) it is quoted as 37.5 kN.

Table 4.1 indicates that there is enough bond resistance available for all strips except 1 and 2. These strips must be either widened or lengthened.

For strip 1 the additional bond resistance due to S_L is:

$$2 \times \frac{100}{1000} \times 1.75 \times 0.29 \times 114.3 = 11.6 \text{ kN}.$$

∴ The additional bond resistance required = 39.8 − (22.2 + 11.6) = 6.0 kN which can be obtained by increasing the reinforcement strip length by ΔL where

$$\Delta L = \frac{6 \times 1000}{2 \times 100 \times 0.29 \times 45} = 2.3 \text{ m}.$$

Hence, total length of strip 1 should be 8.5 + 2.3 = 10.8 m, say 11 m.

The reader might like to check that a length of 11 m gives adequate bond resistance to strip 2.

Hence, from the local stability approach:

Cross sectional area of reinforcing elements = 3 mm × 100 mm

Length of reinforcing elements = 8.5 m

$$S_v = 0.75 \text{ m}; S_H = 0.5 \text{ m}.$$

(ii) Wedge stability

Let the reinforced earth fill be divided into three equal zones (more zones may be necessary in a practical problem) to give the points A, B and C as the points on the wall facing from which various trial failure surfaces may be drawn (fig. 4.20).

The Coulomb failure wedge is inclined at $[45° + (\phi/2)]$ to the horizontal, i.e. at an angle to the vertical, β, of $[45° - (\phi/2)]$. Any group of trial failure surfaces should include the Coulomb wedge with two or more trial surfaces on either side, i.e. suitable angles for β are 20°, 25°, 30°, 35° and 40°.

Note. In this form of analysis it is simpler to regard S_L as a concentrated line load rather than one that is spread through a footing.

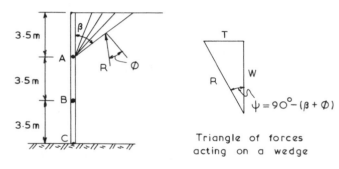

Fig. 4.20 Wedge stability

Consider the wedges drawn from point A in fig. 4.20.

Point A is at 3.5 m below the top of the wall, i.e. reinforcing elements 1 to 4 inclusive lie within all wedges drawn from A. Calculations are best tabulated (table 4.2):

Table 4.2

β	Length of top of wedge (m)	Weight of wedge, $W =$ $\gamma \times$ area (kN)	w_s (kN)	S_L (kN)	Total W (kN)	ψ	T $(= W \tan \psi)$ (kN)
20°	1.27	44.45	38.10	0	83	40°	69
25°	1.63	57.05	48.90	0	106	35°	74
30°	2.02	70.70	60.60	0	131	30°	76
35°	2.45	85.75	73.50	200	359	25°	168
40°	2.94	102.90	88.20	200	391	20°	142

Statistics and soil mechanics 181

histogram is as shown in fig. 5.1C. This is still fairly coarse and a more suitable height increment would appear to be 2.5 cm. This results in the histogram shown in fig. 5.1D where the horizontal scale is split into 16 equal increments.

Table 5.2

Height range	1st column	1st & 2nd col.	All columns
150 - 155		/	/
155 - 160			/
160 - 165	///	///	‖‖‖ ////
165 - 170	//	///	‖‖‖ ‖‖‖ ‖‖‖ ///
170 - 175	////	‖‖‖ ///	‖‖‖ ‖‖‖ ‖‖‖ ‖‖‖ ‖‖‖ ‖‖‖ ////
175 - 180	/	////	‖‖‖ ‖‖‖ ‖‖‖ ‖‖‖ ‖‖‖
180 - 185		/	‖‖‖ ‖‖‖ ///
185 - 190			///

Treatment of values coinciding with boundaries

The convention used in forming the histograms was to assume that any value coinciding with a boundary lies in the increment to the right, i.e. a value of 170.0 cm is assumed to be 170.01 cm. This assumption is generally accurate enough but, for greater refinement, even boundary values can be considered as being in the right-hand increment and odd values as being in the left-hand increment, i.e. 170.0 is assumed to be 170.01 whereas 165.0 is assumed to be 164.99.

A point illustrated by figs. 5.1C and 5.1D is that, for a given set of data, the area under a histogram varies with the width chosen for the blocks.

In fig. 5.1C, total area =
  Width of blocks × total number of observations =
  $5(1 + 1 + 9 + 18 + 34 + 30 + 14 + 3) = 5 \times 110 = 550$
In fig. 5.1D, total area =
  $2.5(1 + 1 + 3 + 5 + 7 + 12 + 16 + 18 + 17 + 13 + 8 + 5 + 3 + 1) = 2.5 \times 110 = 275.$

It is thus seen that the area under a histogram is of little consequence. What is important is that, no matter what width of blocks is decided upon, the total area under the histogram, divided by the block width, gives the total number of observations (in this case 110).

182 *Elements of Foundation Design*

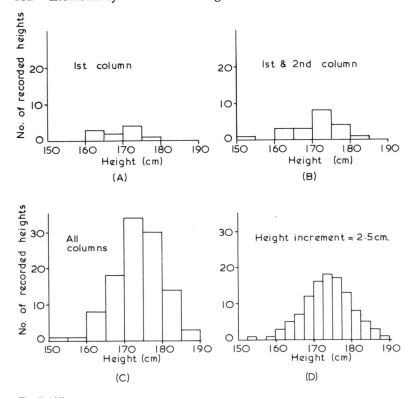

Fig. 5.1 Histograms

Frequency curve

A histogram is a visual display of a set of variables (i.e. measurements) which can show, at a glance, the frequency of occurrence of any particular variable.

The diagram is essentially stepped but can be refined by simply drawing a smooth curve through the mid point of the top of each block. For example, figs. 5.1C and 5.1D yield the curves shown in fig. 5.2A and 5.2B.

Fig. 5.1D, with its smaller width of blocks, tends to yield a smoother curve. One can imagine that if the number of students had been increased ten fold and the accuracy of the measuring equipment was such that the blocks could have been reduced in thickness to say 0.1 cm, an even smoother curve would result.

Obviously there is a theoretical curve that represents the condition of the thickness of the blocks tending to zero as the number of observations tends to infinity. This is known as a frequency curve of which fig. 5.2B is an approximation.

Although a frequency curve represents an idealised theoretical situation involving an infinite number of observations, in practice any reasonable histogram will produce a curve that is a close approximation to the frequency curve of the system.

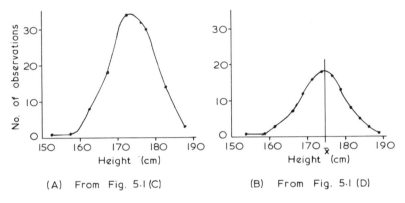

Fig. 5.2 Frequency curves

Average frequency value

Engineers like to think of average values. No-one can work with a set of numbers in which each represents a measurement of some variable but one can work with the average value of the numbers. What is important is that the value chosen as the average must be representative of the set from which it was derived.

An obvious choice for the average of a set of numbers is the arithmetic mean, i.e. the sum of all the numbers or values in the set, divided by the total number of numbers or values.

In symbols: $$\bar{x} = \frac{S}{n}.$$

It may not always be possible to obtain the mean value directly from the set of values as the information available may be in the form of either a histogram or a frequency curve. Obviously the form of grouping of the individual values will induce some loss of accuracy

184 *Elements of Foundation Design*

but, provided there is a realistic number of blocks, the arithmetic mean obtained from a histogram is satisfactory for most statistical purposes.

Example 5.2

The arithmetic mean of all the values in table 5.1 is 173.68.

From fig. 5.1C (taking central values of each block):

$$\bar{x} = \frac{1}{110} (152.5 + 157.5 + 9 \times 162.5 + 18 \times 167.5 + 34 \times 172.5$$
$$+ 30 \times 177.5 + 14 \times 182.5 + 3 \times 187.5)$$
$$= 173.59.$$

From fig. 5.1D:

$$\bar{x} = \frac{1}{110} (153.75 + 158.75 + 3 \times 161.25 + 5 \times 163.75$$
$$+ 7 \times 166.25 + 12 \times 168.75 + 16 \times 171.25 + 18 \times 173.75$$
$$+ 17 \times 176.25 + 13 \times 178.75 + 8 \times 181.25$$
$$+ 5 \times 183.75 + 3 \times 186.25 + 188.75)$$
$$= 173.70.$$

Note. The mean value can also be obtained from a frequency curve. Suitable values are selected and the frequency of these values obtained. The mean value is then determined by averaging out, as for the histogram. In the case of a frequency curve that is symmetrical about its centre (as fig. 5.2B) the mean value is the central value.

Other ways of expressing the average value

Although the arithmetic mean is the most commonly used method for expressing the average of a distribution, particularly in engineering, there are two other terms which can be used, the mode and the median, and the reader should at least be aware of them.

The mode (or modal value)

The word 'mode' means fashion and the modal value of a set of numbers is simply the number most in fashion, i.e. the most commonly occurring value.

The mode, therefore, is the value of variable coresponding to the peak of the frequency curve, or it is the central value of the highest block in a histogram. From the frequency curve of fig. 5.2B, it is seen that the modal value is about 174 cm.

The median

The median is simply the central value of a set. On the one side all the values are less than the median, on the other side all the values are higher.

This means that if all the values in a set of data were placed in order of increasing magnitude then the median is the central value (if the number of values is odd) or is the average of the two central values (if the number of values is even). For the values given in table 5.1:

$$\text{Median} = \frac{v_{55} + v_{56}}{2} = \frac{173.8 + 173.9}{2}$$
$$= 173.85.$$

Standard deviation

As mentioned, the most commonly occurring average in engineering, and the one that this chapter will be concerned with is the arithmetic mean. The deviation of a variable in a set is simply the difference of its value from the mean value.

i.e. deviation of a value = (value − mean value).

For instance, in example 5.1:

Deviation of the value $172.3 = 172.3 - 173.7 = -1.4$
Deviation of the value $176.5 = 176.5 - 173.7 = 2.8$.

Quite often it is necessary in statistics to have a figure that represents the average value of all the deviations (i.e. the mean deviation).

Consider a set of values: 1, 2, 3, 4, 5, 6, 7, 8, 9, 10 and obtain the mean deviation of the set.

$$\text{Arithmetic mean} = \frac{55}{10} = 5.5.$$

To obtain the mean deviation one might think of using a formula such as:

$$\frac{\text{sum of individual deviations}}{\text{number of deviations}}$$

but this would lead nowhere as the sum of the individual deviations is zero when the algebraic signs are taken into account:

Sum of individual deviations =

$-4.5 - 3.5 - 2.5 - 1.5 - 0.5 + 0.5 + 1.5 + 2.5 + 3.5 + 4.5 = 0.$

The value taken as the mean deviation is the root mean square value of the deviations. It is given the symbol σ and is referred to as the standard deviation.

Standard deviation = R.M.S. value of total deviations

$$\text{or } \sigma = \sqrt{\left(\frac{\Sigma(x - \bar{x})^2}{n}\right)}$$

In the above example:

$$\sigma = \sqrt{\left(\frac{(-4.5)^2 + (-3.5)^2 + (-2.5)^2 + \ldots + (3.5)^2 + (4.5)^2}{10}\right)}$$
$$= \sqrt{8.25} = 2.87.$$

The simplicity of the standard deviation formula illustrates the simplicity of the arithmetic involved. It does not indicate the tedium, even with a modern calculator, when the number of values is large.

This tedium can be greatly reduced if the formula is rearranged and written as:

$$\sigma = \sqrt{\left(\frac{\Sigma x^2}{n} - \bar{x}^2\right)}$$

The reader might like to determine σ for the ten values using the new formula. As a further example consider fig. 5.1D and determine σ with the new formula:

$$\sigma = \sqrt{\left(\frac{(153.75^2 + 158.75^2 + 3 \times 161.25^2 + 5 \times 163.75^2 + \ldots \text{etc.})}{110} - 173.7^2\right)}$$
$$= \sqrt{42.145} = 6.49 \text{ cm}.$$

Statistics and soil mechanics

Note. As the reader will appreciate, it is quite possible for a completely different set of data, with a different distribution, to have the same arithmetic mean as another set. A knowledge of the arithmetic mean is therefore of little value unless it is accompanied by the relevant frequency curve or some information, such as the standard deviation.

Coefficient of variation

It is often quite useful to be able to compare the degree of spread of two sets of variables about their respective means. This can be done by a term known as the coefficient of variation, which is simply the standard deviation divided by the mean value, expressed as a percentage.

As the coefficient of variation is dimensionless it can be used to compare sets of variables expressed in different dimensions.

$$\text{Coefficient of variation, } V = \frac{\sigma}{\bar{x}} \times 100\%.$$

Example 5.3

Consider two sets of variables:

I: 11,   13,   18,   24,   28,   32,   35,   40
II: 111, 113, 118, 124, 128, 132, 135, 140

With a little thought it should be possible for the reader to accept that the two sets of values will have the same standard deviation and he should verify that it is equal to 9.83.

Mean of set I = 25.125
Mean of set II = 125.125

$$\therefore V \text{ (for I)} = \frac{9.83}{25.125} \times 100 = 39.12\%$$

$$V \text{ (for II)} = \frac{9.83}{125.125} \times 100 = 7.86\%.$$

Hence, as expected, the spread of values in set I from their mean, is far greater than the spread of the values of set II.

Normal distribution curve

Frequency curves can have various forms but the most common, particularly in engineering problems, is the humped back, or single peaked, distribution with the maximum number of frequencies tending to occur towards the centre of the curve. This bell-shaped distribution curve, already encountered in fig. 5.2B, is very close to a theoretical curve known as the normal distribution curve.

It is represented by the equation,

$$y = \frac{1}{\sigma\sqrt{(2\pi)}} e^{-(x-\bar{x})^2/2\sigma^2}$$

where y is the height of the curve at value x (fig. 5.3A).

Where the maximum number of frequencies is off centre the distribution is termed 'skewed'. Most frequency curves dealing with characteristics of soil, rock and concrete tend to exhibit some skew but, quite often, these curves are close enough to the theoretical normal distribution curve that they can be assumed to possess its characteristics.

It is generally ageed that, provided the coefficient of variation is not greater than 30%, a distribution can be assumed to be normal for most engineering purposes (Hald, 1952). Most soil parameters tend to satisfy this condition and, unless there is definite knowledge to the contrary, it may be assumed that they tend to follow a normal distribution (Lumb, 1974; Singh and Lee, 1972).

Even when a distribution is not close to the normal it can often be transformed to normality by some mathematical manipulation such as plotting the square roots, or the logarithms of the values. An introduction to these techniques is given by Gregory (1978).

The statement that a set of variables is normally distributed means that, if the variables were put into groups to form a histogram, the resulting frequency curve would be symmetrical about the mean value and would have the equation of the normal distribution curve as given above (σ and $\bar{x}$ being calculated from the set of variables).

With a normal distribution the x values of points on the frequency curve are symmetrical about the mean value, $\bar{x}$. Therefore $\bar{x}$ can be subtracted from each x value to give a new horizontal scale $(x - \bar{x})$. With this new scale each measured variable can be plotted by its distance from the mean value, i.e. distance = $(x - \bar{x})$ (fig. 5.3B).

Fig. 5.3B can be converted into a useful dimensionless diagram, applicable to all normal frequency curves, if the horizontal scale

Statistics and soil mechanics 189

$(x - \bar{x})$ is divided by the standard deviation. This gives a new scale, $(x - \bar{x})/\sigma$, and it is then possible to plot each measured variable, x, by its distance (expressed as multiples of the standard deviation) from the mean value $\bar{x}$ (fig. 5.3C).

These values of distance are often referred to as 'standard scores' and given the symbol, z.

i.e. for a particular value of x, $z_x = \dfrac{x - \bar{x}}{\sigma}$.

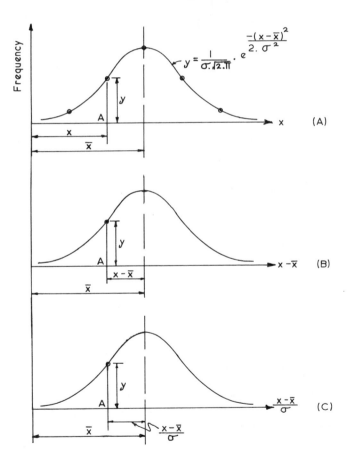

Fig. 5.3 Normal distribution curve

Fig. 5.3C represents the 'standard' normal distribution curve that is widely used in statistics.

If the total area under the curve is regarded as representing the total number of variables and is given the value 1.0 (or 100%) then it becomes possible to determine what percentage of the variables is less (or greater) than variable x (represented by point A in fig. 5.3C). It is the percentage of the total area to the left (or to the right) of A, represented by P_x (or Q_x).

Values of P_x and Q_x for distances σ, 2σ and 3σ, plus another important distance, 1.64σ, are shown in fig. 5.4. Other values of P_x and Q_x can be readily obtained with a programmable calculator.

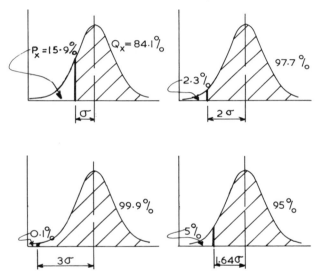

Fig. 5.4 Areas under the normal distribution curve

Sampling techniques

In civil engineering works soil is used in either its undisturbed state or in some remoulded form. In either case it is generally necessary to collect and test representative samples.

Undisturbed soil: Cuttings, foundations, excavations, etc., the purpose of testing is to classify the soil types encountered and to check that their properties, i.e. strength, compressibility, etc., are compatible with the design parameter values.

Remoulded soil: Occurs in earthworks, the placement and compaction (or stabilisation) of soil in the construction of embankments, earth dams, pavements, etc. In these cases tests carried out are mainly control tests to check that the form of the soil finally produced satisfies the requirements of the specification.

The need for a statistical approach

The number of samples tested, where they are taken from, the standard of testing and the decisions on borderline cases often depend largely upon the experience of the inspecting engineer.

The system inevitably leads to subjective judgements of varying standards, varying even with the same engineer, and the risk of a wrong decision is considerable.

It is impossible to guarantee that good material will always be accepted and that bad material will always be rejected.

However, if statistical methods are used in the selection of samples and the interpretation of the test results a system can be evolved that is equally fair to the contractor and to the client. The contractor can be sure that if his product is at least equal to some minimum standard it will generally be accepted and, at the same time, the client knows that if the material is below that standard it will generally be rejected.

Random numbers

A random number is simply a number selected from a group of numbers in such an unbiased way that all the numbers of the set have an equal chance of being selected.

This is not as easy as it sounds. Good approximations to randomness can be obtained by dealing cards or throwing dice but these methods often show some form of bias after a time.

Tables of random numbers have been used in the past but nowadays most programmable calculators have a standard library routine whereby a set of random numbers, between specified limits, can be created in a few minutes.

Selection of samples

For results from a set of samples to be meaningful two main conditions must be satisfied:

192 *Elements of Foundation Design*

The samples must be representative of the material (or population) from which they were taken.

The number of samples must be sufficient.

Selection of representative samples

Whenever possible samples should be selected in an unbiased way, preferably by the use of random numbers.

Example 5.4

It has been decided that ten control tests will be carried out on 100 lorry loads of sand that will be arriving at a construction site over a period of one week. Which loads should be sampled?

Procedure is to select ten random numbers, number the lorry loads consecutively and sample the load whenever the number coincides with a selected random number.

The following ten random numbers were determined using a programmed calculator:

3, 11, 15, 19, 22, 30, 44, 56, 66, 67

and these would be the loads tested.

Example 5.5

It is necessary to determine the chainage at which twenty randomly selected samples should be taken from a 1000 m length of embankment.

The procedure is simply to use the calculator to determine twenty random numbers between 1 and 1000. One such set of numbers is:

22	101	106	147	176
185	192	217	295	437
475	513	558	582	658
667	695	707	873	939

These would be the distances in metres from one end of the embankment at which samples would be taken.

Selection of sample size

Before evolving rules for the determination of the minimum sample size, i.e. the number of samples required, one or two more statistical ideas need to be introduced.

Variance

When considering standard deviation an expression was determined for the value of the average of the squares of the deviations from the mean, viz.:

$$\frac{\Sigma(x - \bar{x})^2}{n} \quad \text{or} \quad \left(\frac{\Sigma x^2}{n} - \bar{x}^2\right)$$

This squared term is important enough to be given a title of its own and it is called the variance of the set of data. The square root of the variance gives the standard deviation.

The frequency curve of fig. 5.3C is reproduced in fig. 5.5. It should be remembered that it was obtained by considering the individual values in a set of data so that the standard deviation, σ, was for a set of individual values i.e. ($n = 1$).

Instead of considering each individual value, the frequency curve may be drawn in three stages:

(a) Grouping the values into fives ($n = 5$) (by random selection).
(b) Determining the mean value of each group.
(c) Using these mean values to obtain $\bar{x}$ and the standard deviation, σ_5.

The value of $\bar{x}$ will, of course, be the same but it will be found that the variance will be much less (fig. 5.5).

If the individual values had been grouped in tens ($n = 10$), $\bar{x}$ would still have the same value but the variance would be even further reduced (fig. 5.5).

In other words, the variance of groups of values is related to n, the number of values in the groups. This relationship is expressed:

Variance of sample with n values/group

$$= \frac{\text{Variance obtained from individual values}}{\text{Number of items per group}}$$

or
$$\text{var}_n = \frac{\sigma^2}{n}.$$

Now, standard deviation is the square root of the variance:

$$\therefore \sigma_n = \frac{\sigma}{\sqrt{n}}.$$

Elements of Foundation Design

Reference to the frequency curve for n = 10 shows that it is symmetrical about the same $\bar{x}$ value as was obtained when considering the individual numbers.

Now, remembering that each value on the n = 10 curve represents a mean of 10 values, it is seen from fig. 5.4 that, for all practical purposes, the mean of any ten samples will not be further from the individual mean value, $\bar{x}$, than $\pm 3\sigma_{10}$, i.e. $\pm 3(\sigma/\sqrt{10})$ (actual probability percentage is 99.7%). Similarly, it is seen that most mean values of any ten values will not be more than $\pm (2\sigma/\sqrt{10})$ from $\bar{x}$ (actual percentage probability = 95.45%).

Hence, using the reverse argument, the true mean value $\bar{x}$ of a population can be established given the mean value $\bar{x}$ of n samples, since:

$$\overline{X} = \bar{x} \pm \frac{3\sigma}{\sqrt{n}} \quad \ldots \text{ with a 99.7\% probability}$$

$$\overline{X} = \bar{x} \pm \frac{2\sigma}{\sqrt{n}} \quad \ldots \text{ with a 95.45\% probability}$$

$$\overline{X} = \bar{x} \pm \frac{\sigma}{\sqrt{n}} \quad \ldots \text{ with a 68\% probability}$$

where $\overline{X}$ = the true mean value of the population
σ = the standard deviation of the total population.

Note. It should be remembered that

$$\frac{3\sigma}{\sqrt{n}} \; ; \quad 2\frac{\sigma}{\sqrt{n}} \quad \frac{\sigma}{\sqrt{n}}$$

represent the limits of the ranges. In most cases $\overline{X}$ will be considerably nearer to $\bar{x}$ than the limiting value.

Standard error of the mean

In the above formula the value controlling the range within which $\overline{X}$ lies is $(\sigma/\sqrt{n})$ and it is called the standard error of the mean (S.E.)

i.e. $$\text{S.E.} = \frac{\sigma}{\sqrt{n}}.$$

Estimation of population standard deviation from sample

With a set of samples it is only possible to determine the sample mean, $\bar{x}$, and the sample standard deviation, s.

In the foregoing formulae one must either use s in place of σ or make an estimate of σ, based on s.

It is now generally accepted that if s is the standard deviation of n samples then the best estimation of σ is given by the formula:

$$\hat{\sigma} = s \sqrt{\left(\frac{n}{n-1}\right)}$$

This correction of s is often referred to as the Bessel correction.

It is seen that the correction is equivalent to changing the formula for the standard deviation from

$$\sqrt{\left(\frac{\Sigma(x-\bar{x})^2}{n}\right)} \quad \text{to} \quad \sqrt{\left(\frac{\Sigma(x-\bar{x})^2}{n-1}\right)}.$$

Some operators only use the formula involving $(n-1)$ as it then becomes unnecessary to check whether or not a correction is required for a low n value. With large n values the difference between the two formulae becomes negligible.

To avoid confusion a reader with a programmable calculator should check which formula his machine uses.

Example 5.6

A series of unconfined compression tests was carried out on samples of a saturated clay. The following c_u values (in kN/m²) were obtained:

94, 98, 109, 92, 107, 106, 99, 98, 104, 105

Assuming a normal distribution, determine the range of c_u values within which the in situ value of the soil deposit will lie. (i) With a probability of 68%; (ii) With a probability of 95.5%; (iii) With a probability of 99.7%.

Mean value = $\bar{x}$ = 101.2 kN/m²

Standard deviation = $s = \sqrt{\left(\frac{\Sigma x^2}{n} - \bar{x}^2\right)} = 5.49$

Using the Bessel correction:

$$\hat{\sigma} = 5.49 \sqrt{\frac{10}{9}} = 5.79$$

With 68% probability $\overline{X} = \bar{x} \pm \dfrac{\hat{\sigma}}{\sqrt{n}} = 101.2 \pm 1.83$

196 *Elements of Foundation Design*

i.e. $\overline{X}$ lies within the range 99.4 to 103.0 kN/m².

With a 95.5% probability $\overline{X}$ lies within the range

$$101.2 \pm 2 \times \frac{5.79}{\sqrt{10}} = 97.5 \text{ to } 104.9 \text{ kN/m}^2.$$

With a 99.7% probability $\overline{X}$ lies within the range

$$101.2 \pm 3 \times \frac{5.79}{\sqrt{10}} = 95.7 \text{ to } 106.7 \text{ kN/m}^2.$$

It is seen that, depending upon the required probability, the range within which $\overline{X}$ lies varies. The greater the probability the greater the range.

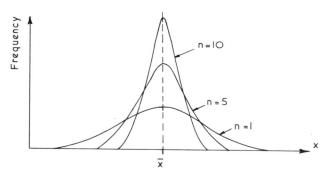

Fig. 5.5 Frequency curves of groups of values

Treatment of small number of samples

When the number of samples is very small the standard formula just discussed cannot be assumed to apply for high probabilities (90% upwards).

For example, for a required 90% probability, the formula should not be taken as $\overline{X} = \bar{x} \pm [2(\hat{\sigma}/\sqrt{n})]$ if the sample number is small. This applies even when it is known that the data are normally distributed.

The expression should be rewritten as:

$$\overline{X} = \bar{x} \pm t\frac{\hat{\sigma}}{\sqrt{n}}$$

where t is obtained from fig. 5.6 (known as Student's t graph).

The procedure is to obtain a value for (n − 1), often referred to as the degrees of freedom, and then to read off the t value corresponding to the required probability.

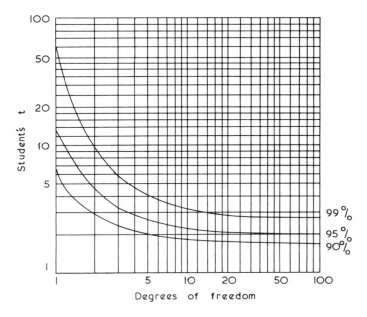

Fig. 5.6 Student's t graph

Characteristic value

CP 110, Part I (1972) defines characteristic strength as the value of the cube strength of the concrete, the yield or proof stress of the reinforcement, or the ultimate load of a prestressing tendon below which not more than 5% of the test results will fall.

A similar definition applies to soil.

For a normal distribution, at a distance ks from the mean, the value will be:

$$\bar{x} - ks$$

where $\bar{x}$ = arithmetic mean of sample results
 s = standard deviation of sample
 k = a variable depending upon the percentage of the sample considered.

From fig. 5.4, it is seen that:

Highest value of lowest 15.9% = $\bar{x} - 1.0s$
Highest value of lowest 2.3% = $\bar{x} - 2.0s$
Highest value of lowest 5% = $\bar{x} - 1.64s$
   i.e. characteristic value = $\bar{x} - 1.64s$.

Example 5.7

Using the c_u values obtained for the saturated clay in example 5.6, determine the characteristic c_u value.

Characteristic c_u value = Mean c_u value $- 1.64 \times$ standard deviation
$= 101.2 - 1.64 \times 5.49 = 92.2 \text{ kN/m}^2$.

Specification and control

For works involving undisturbed soil, characteristic values of the relevant parameters can be obtained in advance of construction and a suitable design prepared.

Soil testing therefore consists of checking that the in situ values are at least of the same order as the design values. Duplication of testing can become expensive and there is obviously scope here for simple in situ tests which are cheap to perform and can be correlated with more expensive laboratory tests. Some work has been carried out in this direction, examples are: the correlation between consistency limits and the compression index of normally consolidated clays; the relationship between standard penetration test results and unconfined compression tests. Much work and research still remains to be carried out before (and if) reliable results are to be obtained.

With remoulded soils control tests must be capable of being carried out relatively quickly so that there is the minimum of delay in construction whenever an acceptance or rejection decision is being decided upon.

The most common tests used in earthworks are:

     Particle size distribution
     Consistency limits
     In situ density
     Moisture content.

The last two tests make it possible to determine the degree of compaction achieved, either in terms of dry density or percentage of air voids.

The acceptance criteria specified should be related to the performance of other earthworks satisfactorily completed in the area. Generally satisfactory dry density values will have a coefficient of variation between 2 and 10%, corresponding to a standard deviation of 3 to 4.5% in the air voids.

Site investigation procedure

The engineering properties of a soil despoit can vary considerably, even when the deposit appears to be homogeneous.

There are two major sources of uncertainty in the determination of values for soil parameters:

The in situ variation of the soil; due to processes involved during formation, subsequent stress history, mineral decomposition, depth of sampling, etc.

Measurement errors, both on the site and in the laboratory; poor operators, sample disturbance, unsuitable or faulty equipment, etc.

As has been discussed, it is necessary to know the mean value and the standard deviation of a set of test results if the characteristic value for a particular parameter is to be obtained. For this value to be dependable a fairly large number of samples is required for each soil type.

The problem is simply that the more samples collected and tested the greater the dependability but the greater the cost.

Site investigations will probably be best carried out in two separate stages.

First there will be an investigation to determine the various different soil deposits present on the site and which of them will need to be sampled. During this part of the investigation as many soil samples as practical should be collected.

From the borehole journals and the sample test results it should be possible to split the sub-strata into a set of sub-regions, each of which can be regarded as being independent of the others (Vanmarke, 1977).

From these preliminary tests an estimate of the minimum number of samples required from each sub-region can be made. If there are not enough samples then the second part of the site investigation becomes necessary so that such samples are obtained.

The minimum number of samples required depends upon various factors, not the least being the accuracy of prediction demanded by the design engineer. If he demands that the average value of the test results should equal the average in situ value he is demanding the impossible as the required number of samples to satisfy this condition is infinite. Fortunately, most engineers will accept a value within 10%.

Example 5.8

Unconfined compression tests carried out on five samples of a stiff clay gave the following undrained shear strengths, c_u, in kN/m²

$$100, \quad 80, \quad 95, \quad 110, \quad 100.$$

Determine: (i) The range of values within which the average in situ value will be 95% certain of lying.

(ii) The total number of samples that will require to be tested if it is specified that the average in situ value is to be within $\pm 10\%$ of the average test value.

(i) Mean c_u test value = 97 kN/m²
$s = 9.8$ kN/m².

Now $\hat{\sigma} = s\sqrt{\left(\dfrac{n}{n-1}\right)} = 9.8 \times \sqrt{\dfrac{5}{4}} = 10.96$ kN/m²

and, as n = 5; S.E. $= \dfrac{\hat{\sigma}}{\sqrt{n}} = \dfrac{10.96}{\sqrt{5}} = 4.9$ kN/m²

$\overline{X} = \bar{x} \pm 2$ S.E. $= 97 \pm 2 \times 4.9 = 87.2$ to 106.8 kN/m².

Note. As already mentioned, Student's t value should really be used for such a small sample. For a 95% probablity and (n − 1) = 4; t = 2.8 (from fig. 5.6).

Hence $\overline{X} = \bar{x} \pm 2.8$ S.E. $= 83.3$ to 110.7 kN/m².

This is the range that should be quoted in answer to part (i) and it is 14% either side of the mean.

(ii) If the range either side of the test mean is to be only $\pm 10\%$, more samples will be required.

Initially, assume that Student's t value is still 2.8.

Then $\overline{X} = \bar{x} \pm 2.8$ S.E. $= \bar{x}(1.0 \pm 0.1)$.

$$\therefore 2.8 \frac{\hat{\sigma}}{\sqrt{n}} = 0.1\bar{x} = 9.7$$

$$\therefore n = \left(\frac{2.8 \times 10.96}{9.7}\right)^2 = 10.$$

The calculation must now be repeated in an iterative manner. From fig. 5.6 the t value corresponding to $(n - 1) = 9$, with a 95% probability, is $t = 2.3$ and n evaluates out at 6.75.

Obviously the true value of t must lie between 2.8 and 2.3. With further iteration it is found that for 95% probability, $n = 8$.
Note. Only when this total number of 8 samples has been tested should $\bar{x}$ and s be obtained, in order to determine the characteristic value of undrained shear strength from the formula $x - 1.64\,s$.

Limit state design

Ultimate limit state of a structure

The ultimate limit state of a structure is its state at collapse.

A short while after being constructed a structure may well be at its greatest strength. With the passage of time there will be a gradual deterioration of the structure and its factor of safety against collapse will slowly reduce until it eventually reaches a value close to 1.0 as the ultimate stress values of its components reduce to the values of the working stresses and collapse is imminent. The structure has reached its ultimate limit state.

Serviceability limit state of a structure

Without doubt the collapse of a structure marks the end of its life but another limit worthy of consideration is the state of the structure which marks the end of its useful life, even though it continues in existence. An orbiting satellite's useful life will depend upon how long the electrical equipment on board continues to function although the satellite may continue in orbit for several years after it has ceased to be of use. The serviceability limit state of a structure is therefore the point at which it ceases to be capable of carrying out the function for which it was designed.

The reader can no doubt think of other situations that would create a serviceability limit state. Some examples are excessive deflections, high levels of vibration, fire damage.

Elements of Foundation Design

Probability theory and design

It is obviously more economical if a structure can be designed to resist its serviceability states rather than a one-off ultimate limit state.

The idea is not new to soil mechanics. For many years foundation engineers have concurrently considered the problem of ultimate limit state (bearing capacity failure) and serviceability state (excessive settlement).

With this approach a partial factor of safety is applied to every quantity, be it load or strength of material, which is not known accurately. Generally these factors increase the calculated loads, by multiplication, and decrease the assumed strengths, by division.

Brinch Hansen (1970) described how the technique was used in Denmark for soil mechanics problems. The partial factors of safety he quoted were:

For cohesion	1.75
For $\tan \phi$	1.25
For superimposed loads	1.5

Eventually the day will come when both the applied loads and the strengths of the various materials involved in a structure will be considered as random variables, capable of being analysed statistically.

In simple terms, if a structure with a deterministic strength, R (fig. 5.7B) is subjected to a random load, S (fig. 5.7A), failure will occur when $S > R$.

The probabilistic design method therefore consists of determining the probability of the event $(S > R)$, i.e. the probability of failure.

If this probability is deemed acceptable from the points of view of loss of life and economics, then the design is considered acceptable.

In other words, the fundamental premise of probability design is that there is no such thing as a 'safe' structure. There will always be a risk of failure, a risk that can never be eliminated.

The probability of failure will occur when the load exceeds the strength and this can be obtained by combining figs. 5.7A and B. The probability of failure is indicated where these figures overlap, shown shaded in fig. 5.7C.

By subtracting fig. 5.7B from fig. 5.7A the probability curve of strength and this can be obtained by combining figs. 5.7A and B.

The partial factors of safety that will be applied to all indeterminate quantities, load or strength, must also be determined by

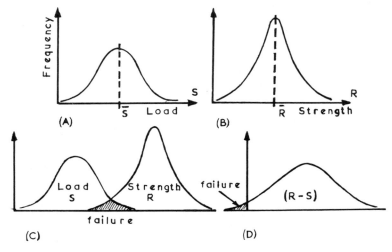

Fig. 5.7 Probability curves of failure

theories of probability and Report No. 63 by C.I.R.I.A. (1976) describes suitable procedures.

Eventually the probabilistic method should cover the whole gambit of the life of a structure. By statistical means it will not only deal with the technical features connected with structural failure, but also with the economics necessary to achieve full maximisation of the use of the structure.

The adoption of probability theories into limit state design is not a situation that will occur overnight but will involve a gradual transition, possibly lasting over several years.

There is already a move towards this changeover, which is evident when the contents of CP 110, Part I (1972) are examined. Probability theories do not feature in the Code which, whilst acknowledging that strengths and loads are variable quantities, accepts that at this stage at least, they must be regaded as constants, i.e. characteristic strengths and loads to which appropriate partial factors of safety must be applied in order to determine suitable design values. The values for the partial factors of safety are quoted as fixed values for the specific situations that arise in design considerations.

Eventually probability theories may feature in future codes of practice but, for the present, the most that the reader will require is a knowledge of sampling techniques and characteristic values.

References (chapter 5)

British Standards Institution (1972) *Code of Practice CP 110: Part 1: The Structural Use of Concrete*. London: B.S.I.

C.I.R.I.A. (Oct. 1976) *Rationalisation of Safety and Serviceability Factors in Structural Codes*. Report No. 63. London: Construction Industry Research and Information Association.

Gregory, S. (1978) *Statistical Methods and the Geographer*. 4th edition. London: Longman Group.

Hald, A. (1952) *Statistical Theory with Engineering Applications*. Chicester: John Wiley.

Hansen, J. B. (1970) 'A revised and extended formula for bearing capacity.' *Bulletin No. 28*. Copenhagen: Danish Geotechnical Institute.

Lumb, P. (1974) 'Applications of statistics in soil mechanics.' *Soil Mechanics–New Horizons*, chapter 3. London: Newnes-Butterworths.

Singh, A. and Lee, K. L. (1972) 'Variability of soil parameters.' *Proc. 8th Annual Symposium on Engng Geology and Soils Engng*. Idaho State University.

Vanmarcke, E. H. (1977) 'Probabilistic modelling of soil profiles.' *J. Geotech. Engng Div. Am. Soc. Civ. Engrs (GT11)*.

Appendix I
Elements of Matrix Algebra

A matrix is simply a collection of numbers written in a series of equal rows:

$$\begin{bmatrix} a & b & c \\ d & e & f \\ g & h & i \end{bmatrix} \quad \begin{bmatrix} 1 & 2 & 3 & 4 \\ 5 & 6 & 7 & 8 \end{bmatrix}$$

The figures are enclosed within square brackets to indicate that they form a matrix. The horizontal lines are known as rows and the number of rows is given the symbol 'm'.

The vertical lines are known as columns and the number of columns is given the symbol 'n'.

In the left hand matrix above m = 3: n = 3.
In the right hand matrix above m = 2: n = 4.

Size of a matrix

The size of a matrix is expressed as m × n. The left hand matrix would be known as a 3 × 3 matrix (usually expressed as a three by three matrix). The right hand matrix is a 2 × 4 matrix.

Square matrix

When m = n the matrix is known as a square matrix. The left hand matrix is a square matrix.

Symbols for matrices

To save writing out all the figures each time, a matrix can be given a symbol, generally a capital letter:

$$A = \begin{bmatrix} a & b & c \\ d & e & f \\ g & h & i \end{bmatrix} \quad C = \begin{bmatrix} 1 & 2 & 3 & 4 \\ 5 & 6 & 7 & 8 \end{bmatrix}$$

Note that when using symbols the square brackets can be dropped.

Elements of a matrix

Each term in a matrix is known as an element and is usually given the symbol 'a' with two suffices to indicate the row and the column in which it lies, e.g. $a_{3,5}$ indicates the element in the third row (counting down from the top) and in the fifth column (counting from the left); e.g. the figure 7 in the right hand matrix would be given the symbol $a_{2,3}$. In general $a_{i,j}$ represents the element in row i and column j.

Vectors

If a matrix consists of only one row or column it is known as a vector and its symbol is a small letter, rather than a capital:

$$x = [x_1 \; x_2 \; x_3 \; x_4] \quad \text{or} \quad d = \begin{bmatrix} 1.2 \\ 3.7 \\ 5.6 \\ 9.0 \end{bmatrix}$$

A vector can be written as either a row or a column, a convenient dodge in mathematical operations.

Multiplication of matrices

In many engineering problems one set of values may be related to another. It may be that there are two vectors, $x = [x_1 \; x_2 \; x_3 \; x_4]$ and $y = [y_1 \; y_2 \; y_3 \; y_4]$ which are related to each other in the form:

$$x_1 = a_{11} y_1 + a_{12} y_2 + a_{13} y_3 + a_{14} y_4 \quad (1)$$

$$x_2 = a_{21} y_1 + a_{22} y_2 + a_{23} y_3 + a_{24} y_4 \quad (2)$$

$$x_3 = a_{31} y_1 + a_{32} y_2 + a_{33} y_3 + a_{34} y_4 \quad (3)$$

$$x_4 = a_{41} y_1 + a_{42} y_2 + a_{43} y_3 + a_{44} y_4. \quad (4)$$

The relationship can be written in matrix form:

$$x = Ay \quad (I)$$

where $x = \begin{bmatrix} x_1 \\ x_2 \\ x_3 \\ x_4 \end{bmatrix} \quad y = \begin{bmatrix} y_1 \\ y_2 \\ y_3 \\ y_4 \end{bmatrix} \quad A = \begin{bmatrix} a_{11} & a_{12} & a_{13} & a_{14} \\ a_{21} & a_{22} & a_{23} & a_{24} \\ a_{31} & a_{32} & a_{33} & a_{34} \\ a_{41} & a_{42} & a_{43} & a_{44} \end{bmatrix}$

The full form of equation (I) would be:

$$\begin{bmatrix} x_1 \\ x_2 \\ x_3 \\ x_4 \end{bmatrix} = \begin{bmatrix} a_{11} & a_{12} & a_{13} & a_{14} \\ a_{21} & a_{22} & a_{23} & a_{24} \\ a_{31} & a_{32} & a_{33} & a_{34} \\ a_{41} & a_{42} & a_{43} & a_{44} \end{bmatrix} \begin{bmatrix} y_1 \\ y_2 \\ y_3 \\ y_4 \end{bmatrix}$$

and the expression means that vector x is obtained by multiplying vector y by matrix A.

For example, equation (1) is obtained by multiplying vector y by the first row of the matrix. This is achieved by placing the row of the matrix alongside the vector and multiplying the corresponding elements together.

$$x_1 = \begin{bmatrix} a_{11} & a_{12} & a_{13} & a_{14} \end{bmatrix} \begin{bmatrix} y_1 \\ y_2 \\ y_3 \\ y_4 \end{bmatrix}$$

A good way to remember this technique is to think of a swimmer diving off a board. A horizontal move along and then the vertical dive:

Once this move has been accomplished the two sets of elements lie alongside each other:

$$x_1 = \begin{bmatrix} y_1 \\ y_2 \\ y_3 \\ y_4 \end{bmatrix} \begin{bmatrix} a_{11} \\ a_{12} \\ a_{13} \\ a_{14} \end{bmatrix}$$

When these corresponding terms are multiplied together we obtain equation (1)

$$x_1 = a_{11}y_1 + a_{12}y_2 + a_{13}y_3 + a_{14}y_4.$$

Hence it is seen that the *first* term of vector x is obtained by multiplying vector y by the *first* row of matrix A.

The reader might like to check that to obtain the *second* term of vector x, vector y must be multiplied by the *second* row of matrix A, and so on.

It should be noted that, for a vector to be multiplied by a matrix there must be the same number of columns in the matrix as there are elements in the vector. Similarly, for a matrix to be multiplied by a matrix there must be the same number of columns in the multiplier matrix as there are rows in the multiplied matrix.

For example:

$$\begin{bmatrix} 1 & 2 & 3 \end{bmatrix} \begin{bmatrix} 2 \\ 4 \\ 6 \\ 7 \end{bmatrix} \text{ is meaningless but } \begin{bmatrix} 1 & 2 & 3 \end{bmatrix} \begin{bmatrix} 2 \\ 4 \\ 6 \end{bmatrix} = 28.$$

Example

$$\begin{bmatrix} 4 & 3 & 2 \\ 6 & 0 & 5 \end{bmatrix} \begin{bmatrix} 6 & 3 & 1 \\ 2 & 9 & 0 \\ 3 & 8 & 1 \end{bmatrix} = A \quad \text{Determine A.}$$

Is multiplication possible? Yes because the multiplier matrix (the one on the left as the horizontal movement is always to the right) has *three* columns whilst the matrix to be multiplied has *three* rows.

What size will matrix A be? Obviously it will have three columns and two rows. It will be a 2×3 matrix.

Let us determine a_{11} of matrix A. To do this we multiply the *first* column by the first row of the multiplier: the row [4 3 2] moves to the right and dives so that the two sets of figures lie alongside each other:

$$a_{11} = \begin{matrix} 6 & 4 \\ 2 & 3 \\ 3 & 2 \end{matrix}$$

When the terms are multiplied together and then added up we obtain a_{11} of matrix A.

$$a_{11} = 6 \times 4 + 2 \times 3 + 3 \times 2 = 36.$$

Similarly, in order to obtain a_{12} of matrix A, i.e. the *second* term of the *first* row we must multiply the *second* column by the *first* row: the *first* row:

$$a_{12} = \begin{bmatrix} 3 \\ 9 \\ 8 \end{bmatrix} \begin{bmatrix} 4 \\ 3 \\ 2 \end{bmatrix} = 55.$$

And, in order to obtain a_{13} we must multiply the third column by the first row. As established at the start of the example, there are only two rows in the product matrix A. The terms of the *second row* are obtained by multiplying each column of the right hand matrix by the *second row* of the left hand matrix.

On multiplying out it is found that:

$$A = \begin{bmatrix} 36 & 55 & 6 \\ 51 & 58 & 11 \end{bmatrix}$$

Unit matrix

A special form of a square matrix is one in which all terms are zero except those on the leading diagonal which are all equal to unity. (A leading diagonal can only occur in a square matrix and is the line joining the top left to the bottom right corners.) Examples of unit matrices are:

$$\begin{bmatrix} 1 & 0 \\ 0 & 1 \end{bmatrix} \quad \begin{bmatrix} 1 & 0 & 0 \\ 0 & 1 & 0 \\ 0 & 0 & 1 \end{bmatrix} \quad \begin{bmatrix} 1 & 0 & 0 & 0 & 0 \\ 0 & 1 & 0 & 0 & 0 \\ 0 & 0 & 1 & 0 & 0 \\ 0 & 0 & 0 & 1 & 0 \\ 0 & 0 & 0 & 0 & 1 \end{bmatrix}$$

The multiplication of a matrix by a unit matrix gives an important result. For example:

$$\begin{bmatrix} 1 & 0 & 0 \\ 0 & 1 & 0 \\ 0 & 0 & 1 \end{bmatrix} \begin{bmatrix} 7 & 8 \\ 4 & 3 \\ 5 & 8 \end{bmatrix} = A.$$

If the rules of matrix multiplication are applied it will be found that the product matrix

$$A = \begin{bmatrix} 7 & 8 \\ 4 & 3 \\ 5 & 8 \end{bmatrix}$$

In other words the original matrix has not been changed by being multiplied by a unit matrix.

This is an important law: multiplication of a matrix, or a vector, by a unit matrix leaves the original matrix, or vector, unaltered.

A unit matrix is given the symbol I.

Multiplication of a matrix by a constant

An expression such as

$$A = 0.25 \begin{bmatrix} 2 & 3 & 4 \\ 6 & 8 & 0 \end{bmatrix}$$

simply means that every element in the matrix must be multiplied by 0.25, giving

$$A = \begin{bmatrix} 0.5 & 0.75 & 1.0 \\ 1.5 & 2.0 & 0 \end{bmatrix}$$

This can be very useful at times. For instance, the rather unwieldy matrix:

$$\begin{bmatrix} 3750 & 6250 & 7500 \\ 5000 & 15000 & 0 \\ 2500 & 5000 & 1250 \end{bmatrix}$$

might be simpler to handle if expressed as:

$$1250 \begin{bmatrix} 3 & 5 & 6 \\ 4 & 12 & 0 \\ 2 & 4 & 1 \end{bmatrix}$$

Inverse matrix

If there is an expression $x = Ay$ which relates vector x to vector y one would expect that there must be a reverse relationship $y = A^{-1} x$.

In the type of engineering problem discussed in this book such a reverse relationship does exist. A^{-1} is called the inverse of matrix A.

We therefore have two expressions:

$$x = Ay \qquad (1)$$
and
$$y = A^{-1}x. \qquad (2)$$

Substituting for y in equation (2):

$$x = A A^{-1} x.$$

This means that the product $A A^{-1}$ must yield a matrix which does not alter x when the two are multiplied together.

The only matrix with this property is the unit matrix.

Hence
$$A A^{-1} = I.$$

There are various ways in which an inverse can be evaluated. For the readers of this book the simplest way is to use a programmable calculator. However, once the inverse has been evaluated, it is well worth while to check that $A\,A^{-1} = I$. This does not mean that the two matrices should be multiplied out completely, merely that one or two spot checks be made, for instance that $a_{22} = 1.0$ and that $a_{21} = 0$. If the reader does any work along these lines he will soon learn to choose the rows and columns with the most zero terms and thereby reduce the effort involved.

Appendix II Flexibility and Rigidity

Anyone undertaking a study of foundation engineering immediately encounters the terms, 'flexible foundation' and 'rigid foundation'. Just what these two terms mean is set out below.

From ordinary beam theory it can be shown that:

$$\frac{M}{I} = \frac{E}{R}$$

where M = bending moment
 E = modulus of elasticity of the material comprising the beam
 I = moment of inertia of the beam
 R = resulting radius of curvature.

Consider a uniformly loaded, simply supported beam, with a central deflection y_c (fig. II.1). The left hand support will be considered as the origin, 0, and the positive x and y directions are as shown.

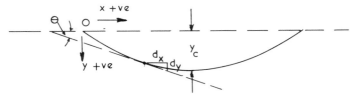

Fig. II.1

The slope of the beam, θ, at any point along it, is defined as the angle made by the tangent at that point to the x axis. For small angles, such as occur in beams, it may be assumed that $\theta = \tan \theta = dy/dx$.

Consider two points, A and B, on the beam at distance dx apart (fig. II.2).

Let the curved length of the beam between A and B = dS.
Let θ = angle to x axis made by the tangent at A.
Then $\theta - d\theta$ = angle to x axis made by tangent at B. (This is because a positive increase in dS causes a decrease in the angle of slope, equal to $- d\theta$.)

214 *Elements of Foundation Design*

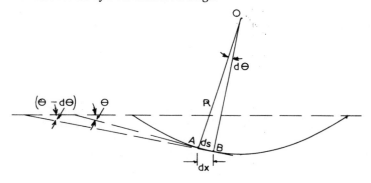

Fig. II.2

Hence $d\theta/dS$ is $-$ve and $1/R = -d\theta/dS$.

In the limit $dS = dx$.

$$\therefore \frac{1}{R} = -\frac{d\theta}{dS} = -\frac{d\theta}{dx} = -\frac{d\left(\frac{dy}{dx}\right)}{dx^2}$$

$$\therefore -\frac{M}{EI} = \frac{d^2y}{dx^2}.$$

Now $\theta = dy/dx = \dfrac{1}{EI} \displaystyle\int -M\,dx = \dfrac{1}{EI}\left[-Mx + A\right]$

and $y = \dfrac{1}{EI} \displaystyle\iint -M\,dx\,dx = \dfrac{1}{EI}\left[\dfrac{-Mx^2}{2} + Ax + B\right].$

For a perfectly rigid beam $y = 0$ and EI must be infinite.
For a perfectly flexible beam $y = \infty$ and EI must be zero.

i.e. If $EI = 0$ the beam is perfectly flexible.
   If $EI = \infty$ the beam is perfectly rigid.

Appendix III
The Finite Difference Method

The differential equation $- [(M/EI) = (d^2y/dx^2)]$ can be solved in a variety of ways. The method that will be described here is that of finite differences which is not only simple to apply but is readily adaptable to the solution of foundation problems.

With the method certain approximations must be made. The first is that the beam to be analysed must be assumed to be split into a number of sections of equal length. For ordinary beam problems the number of these sections can be as little as four but for foundation beams the number should not be less than eight and even more if greater accuracy is required (sometimes necessary if there is a possibility of uplift along the beam).

Assume that the beam has been split into four equal sections and that the deflections at the five nodal points 1, 2, 3, 4 and 5 under some loading system are y_1, y_2, y_3, y_4 and y_5, as shown in fig. III.1.

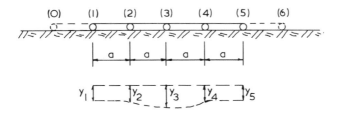

Fig. III.1 Beam split into four sections

Consider first the interior node, point 2:

In the direction 1–2 $dy/dx \approx \Delta y/\Delta x = \dfrac{y_1 - y_2}{a}$.

In the direction 2–3 $dy/dx \approx \Delta y/\Delta x = \dfrac{y_2 - y_3}{a}$.

216 *Elements of Foundation Design*

Now $d^2y/dx^2 \approx \dfrac{\Delta^2 y}{\Delta x^2} = \dfrac{\Delta \left(\dfrac{\Delta y}{\Delta x}\right)}{\Delta x}$

$= \dfrac{1}{a}\left(\dfrac{y_1 - y_2}{a} - \dfrac{y_2 - y_3}{a}\right) = \dfrac{y_1 - 2y_2 + y_2}{a^2}.$

Similar expressions can be derived for points 3 and 4.

Consider now the exterior nodes, points 1 and 5.

There are various mathematical approximations that can be used for d^2y/dx^2 at these external nodes. Possibly the simplest technique is to imagine that extra points exist at a distance a from each edge of the beam (points 0 to 6 in fig. III.1).

With the imaginary point 0 it is now possible to regard point 1 as internal and the finite difference expression for d^2y/dx^2 for point 1 is

$$\dfrac{y_0 - 2y_1 + y_2}{a^2}.$$

Hence the relationships between bending moment values and deflections along the beam can be expressed in matrix form:

$$\begin{bmatrix} M_1 \\ M_2 \\ M_3 \\ M_4 \\ M_5 \end{bmatrix} = -\dfrac{EI}{a^2} \begin{bmatrix} 1 & -2 & 1 & 0 & 0 & 0 & 0 \\ 0 & 1 & -2 & 1 & 0 & 0 & 0 \\ 0 & 0 & 1 & -2 & 1 & 0 & 0 \\ 0 & 0 & 1 & -2 & 1 & 0 & 0 \\ 0 & 0 & 0 & 1 & -2 & 1 & 0 \\ 0 & 0 & 0 & 0 & 1 & -2 & 1 \end{bmatrix} \begin{bmatrix} y_0 \\ y_1 \\ y_2 \\ y_3 \\ y_4 \\ y_5 \\ y_6 \end{bmatrix}$$

y_0 and y_6 are imaginary and must therefore be removed from the equation. This can be accomplished by removing the first and last columns of the matrix to give the final expression:

$$\begin{bmatrix} M_1 \\ M_2 \\ M_3 \\ M_4 \\ M_5 \end{bmatrix} = -\dfrac{EI}{a^2} \begin{bmatrix} -2 & 1 & 0 & 0 & 0 \\ 1 & -2 & -1 & 0 & 0 \\ 0 & 1 & -2 & -1 & 0 \\ 0 & 0 & 1 & -2 & 1 \\ 0 & 0 & 0 & 1 & -2 \end{bmatrix} \begin{bmatrix} y_1 \\ y_2 \\ y_3 \\ y_4 \\ y_5 \end{bmatrix}$$

Example III.1

A simply supported beam of uniform EI carries a central load W. Assuming that the beam is split into four equal sections determine an expression for the central deflection of the beam. Assume that supports 1 and 5 do not yield. The beam is illustrated in fig. III.2.

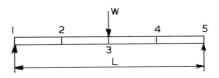

Fig. III.2

The two reactions are equal to $W/2$ and $y_1 = y_5 = 0$.

$$-M_2 = -\frac{W}{2} \times a = -\frac{Wa}{2} = EI\left[\frac{y_1 - 2y_2 + y_3}{a^2}\right] \quad (1)$$

$$-M_3 = -\frac{W}{2} \times 2a = -Wa = EI\left[\frac{y_2 - 2y_3 + y_4}{a^2}\right] \quad (2)$$

$$-M_4 = -\frac{W}{2} \times 3a \times Wa = -\frac{Wa}{2} = EI\left[\frac{y_3 - 2y_4 + y_5}{a^2}\right] \quad (3)$$

But $y_1 = y_5 = 0$ ∴ equations can be rewritten:

$$-2y_2 + y_3 = -\frac{Wa^3}{2EI} \quad (1)$$

$$y_2 - 2y_3 + y_4 = -\frac{Wa^3}{EI} \quad (2)$$

$$y_3 - 2y_4 = -\frac{Wa^3}{2EI} \quad (3)$$

From (1), $y_2 = \dfrac{Wa^3}{4EI} + \dfrac{y_3}{2}$ and from (3), $y_4 = \dfrac{Wa^3}{2EI} + \dfrac{y_3}{2}$.

Substituting in (2) $\dfrac{Wa^3}{4EI} + \dfrac{y_3}{2} - 2y_3 + \dfrac{Wa^3}{4EI} + \dfrac{y_3}{2} = -\dfrac{Wa^3}{EI}$

i.e. $y_3 = \dfrac{3}{2}\dfrac{Wa^3}{EI}$

But a = L/4 $\therefore y_3 = \dfrac{3}{128} \dfrac{WL^3}{EI}$.

Note: Theoretical solution is $y_3 = \dfrac{WL^3}{48EI}$.

Example III.2
A simply supported beam of uniform cross section has a span of 16 m and is subjected to a clockwise moment of 16 kNm at the point shown in fig. III.3.

Considering bending effects only determine an expression, by finite differences, for the deflection of the beam at the point of application of the moment. It can be assumed that the supports of the beam are unyielding.

First find the reactions at the supports and then determine the bending moment diagram.

Taking moments about the left hand support:

$Q_5 \times 16 - 16 = 0$ $\therefore Q_5 = 1$ kN upwards.

Taking moments about right hand support:

$Q_1 \times 16 + 16 = 0$ $\therefore Q_1 = 1$ kN downwards.

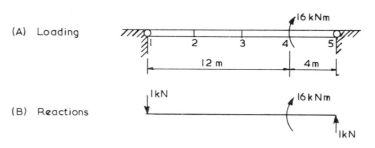

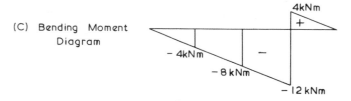

Fig. III.3

The bending moment diagram can now be obtained and values assigned to each nodal point:

$M_2 = -4 \text{ kNm}; \quad M_3 = -8 \text{ kNm}; \quad M_4 = \frac{1}{2}(4 - 12) = -4 \text{ kNm}$

$$-M_2 = 4 = EI \frac{y_1 - 2y_2 + y_3}{a^2} \tag{1}$$

$$-M_3 = 8 = EI \frac{y_2 - 2y_3 + y_4}{a^2} \tag{2}$$

$$-M_4 = 4 = EI \frac{y_3 - 2y_4 + y_5}{a^2} \tag{3}$$

But $\qquad y_1 = y_5 = 0.$

∴ From (1) $\quad y_2 = -\dfrac{2a^2}{EI} + \dfrac{y_3}{2}$

From (3) $\quad y_3 = \dfrac{4a^2}{EI} + 2y_4.$

Substituting in (2) for y_3:

$$-\frac{2a^2}{EI} + \frac{y_3}{2} - 2y_3 + y_4 = \frac{8a^2}{EI}$$

$$\therefore -\frac{2a^2}{EI} - \frac{3y_3}{2} + y_4 = \frac{8a^2}{EI}$$

and substituting for y_3:

$$-\frac{2a^2}{EI} - \frac{3}{2} \times \frac{4a^2}{EI} - \frac{3 \times 2y_4}{2} + y_4 = \frac{8a^2}{EI}$$

$$\therefore y_4 = -\frac{8a^2}{EI} - \frac{8 \times 4^2}{EI} = -\frac{128}{EI} \text{ (an upward deflection)}.$$

Note: sign convention for moments and deflections is given in chapter 3.

Index

Adhesion between pile and soil, 57–60
Adhesion between reinforcement and soil, 156, 157
Allowable bearing pressure, 1
Allowable pile load, 56, 72, 73
Annular cracking around top of pile, 60
Arithmetic mean, 183
Average value, 183–5

Battered piles 95, 96
Bearing capacity
  allowable, 1
  coefficient N_c, 3, 4, 32, 57
  coefficient N_q, 3, 32
  coefficient N_γ, 3, 33
  equations, 2, 6, 8, 9, 36, 37
  maximum safe, 1, 4, 22
  net, 4
  offshore structures, 32–8
  ultimate, of a foundation, 1, 4
  ultimate, of a pile, 55
Bond failure of reinforcing strip, 156, 157
Bored piles, 61
Buoyant foundation 27

Characteristic value, 197, 198
Coefficient of horizontal subgrade reaction, 132, 133
Coefficient of subgrade reaction, 110–14
Coefficient of variation, 187
Cohesive fill, 142
Cohesive frictional fill, 142, 175, 176
Compensated foundations, 27
Consolidation factor, 26
Consolidation factor values, 27
Consolidation settlement, 26
Contact pressure distribution, standard solutions, 101–4
Corrosion, 176, 177
Coulomb theory of active earth pressure, 147–53
Critical depth, 28, 66
Cyclic loading effects, 44–8

Deep foundation, 1
Displacement of pile top, 87–92
Dragdown, 60
Driven piles, 60

Dutch one penetration test, 14–19, 73

Eccentric loading on pile groups, 93–5
Effective foundation, 34–6
Efficiency of pile groups, 79
Elastic foundation, 105
Elastic subgrade, 106
End resistance of piles, 65, 68

Facing units, 142, 172, 173
Finite difference method, 215–19
Flexibility, 213, 214
Flexible foundations, 102
Foundations
  buoyant, 27
  combined, 104–30
  deep, 1
  effective, 34–6
  flexible, 102
  offshore, 29–52
  pad, 2
  piled, 2
  raft, 2
  rigid, 103
  shallow, 1
  strip, 2
Frequency curve, 182, 183
Frictional fill, 141, 174, 175

Histogram, 179–82

Immediate settlement, 22
Inclined load factors, 36
Inclined loads on pile groups, 93–5
Inverse matrix, 210

Lateral loading on piles, 85–96, 131–8
Length of reinforcing elements, 165
Limit state design, 201
Line load, 158, 159

Matrix, 205
Maximum safe bearing capacity, 1
Maximum tension line, 154, 155
Median, 185
Mode, 184
Modulus of Elasticity, 26

Modulus of horizontal subgrade reaction, 132, 133
Modulus of subgrade reaction, 110–14
Multiplication of matrices, 206

Negative skin friction, 74–6
Normal distribution curve, 188–90
Numerical methods, 99–138

Overburden pressure, 1

Pad foundation, 2
Penetration tests, 10–19
Piled foundations, 2
Piles
  allowable load, 56, 72, 73
  bored, 61, 68
  driven, 60, 65
  end bearing, 55, 61
  friction, 55, 68
  groups, 79
  tension, 76
Plate loading tests, 21, 22, 113–15
Poisson's ratio, typical soil values, 23
Pressuremeter, 19, 20
Probability theory, 202–4

Raft foundation, 22–7
Raking piles, 95, 96
Random numbers, 191
Rankine theory of active earth pressure, 147–54
Reinforced earth, 141–77
Reinforcing elements, 142, 151, 155–7, 165
Rigid foundations, 101, 103
Rigidity, 213, 214
Rigid method, 105

Sampling techniques, 190–2
Scour, 51, 52
Serviceability limit state, 201
Settlement
  cohesive soils, 22–7
  granular soils, 9–19
  of pile groups, 82
Shallow foundation, 1
Shape factors, 34
Simplified elastic subgrade, 105, 106
Site investigation procedure, 199, 200
Skin friction, 57–60
Skirting, 37, 38
Smear effects due to dragdown, 60
Specification and control, 198, 199
Square matrix, 205
Standard deviation, 185–7, 194, 195
Standard error of the mean, 194
Standard penetration test, 10, 72, 114
Standard scores, 188, 189
Statistics, 179–203
Strip foundation, 2

Tension line, maximum, 154, 155
Tension piles, 76

Ultimate bearing capacity of a foundation, 1
Ultimate bearing capacity of a pile, 55
Ultimate bearing capacity of a pile group, 80, 81
Ultimate horizontal load on a pile, 86
Unit matrix, 209

Variance, 193
Vectors, 206

Wave action, 30

Civil Engineering books from Granada:

The Constrado Monographs
Each book in the series deals with specific applications of steel in construction and is aimed at practising civil and structural engineers, students, and teaching and research staff in civil engineering at universities.

THIN PLATE DESIGN FOR TRANSVERSE LOADING
B. Aalami and D. G. Williams
0 258 96991 1   212 pages Illustrated

THIN PLATE DESIGN FOR IN-PLANE LOADING
D. G. Williams and B. Aalami
0 246 11236 0   224 pages Illustrated

COMPOSITE STRUCTURES OF STEEL AND CONCRETE: VOLUME 1
R. P. Johnson
0 258 96993 8   228 pages Illustrated

COMPOSITE STRUCTURES OF STEEL AND CONCRETE: VOLUME 2
R. P. Johnson and R. J. Buckby
0 246 11377 4   544 pages Illustrated

DESIGN FOR STRUCTURAL STABILITY
P. A. Kirby and D. A. Nethercot
0 246 11444 4   192 pages Illustrated

STEEL DESIGNERS' MANUAL
Fourth Edition
0 258 96825 7 hardback
0 258 96977 6 paperback
2002 pages Illustrated

MULTI-STOREY BUILDINGS IN STEEL
Hart, Henn, Sontag, Godfrey
0 258 96974 1   360 pages Illustrated

CABLE-STAYED BRIDGES
M. S. Troitsky
0 258 97034 0   400 pages Illustrated

THIN-WALLED STRUCTURES
Recent Trends in Design, Research and Construction
Edited by J. Rhodes and A. C. Walker
0 246 11182 8   808 pages Illustrated

WELDED JOINT DESIGN
J. C. Hicks
0 258 97099 5   96 pages Illustrated

THE FINITE ELEMENT METHOD
A Basic Introduction for Engineers
K. C. Rockey, H. R. Evans, D. W. Griffiths and D. A. Nethercot
0 258 97181 9   252 pages

ADVANCED STRUCTURAL ANALYSIS
Worked Examples
J. Walter White
0 258 97030 8   272 pages Illustrated

REINFORCED CONCRETE DETAILER'S MANUAL
Third Edition
Brian Boughton
0 246 11336 7 paperback
136 pages Illustrated

DESIGN OF REINFORCED CONCRETE ELEMENTS
P. J. B. Morrell
0 258 97018 9 paperback
230 pages Illustrated

TIMBER DESIGNERS' MANUAL
J. A. Baird and E. Carl Ozelton
0 258 97028 6   536 pages Illustrated

ELEMENTS OF SOIL MECHANICS
Fourth Edition
G. N. Smith
0 258 97105 3 hardback
0 246 11334 0 paperback
440 pages Illustrated

ELEMENTS OF FOUNDATION DESIGN
G. N. Smith and E. L. Pole
0 246 11429 0 hardback
0 246 11215 8 paperback
200 pages Illustrated

CASE HISTORIES IN ENGINEERING GEOLOGY
J. G. C. Anderson and C. F. Trigg
0 236 40049 5   212 pages Illustrated

LABORATORY WORK IN CIVIL ENGINEERING: SOIL MECHANICS
Brian Vickers
0 258 97084 7 paperback
160 pages Illustrated

LABORATORY WORK IN HYDRAULICS
W. R. Lomax and A. J. Saul
0 258 97088 X paperback
288 pages Illustrated

SETTING OUT
A Guide for Site Engineers
S. G. Brighty
0 258 96929 6   272 pages Illustrated

MODERN CONSTRUCTION MANAGEMENT
Frank Harris and Ronald McCaffer
0 258 97064 2 hardback
0 258 97164 9 paperback
370 pages Illustrated

WORKED EXAMPLES IN CONSTRUCTION MANAGEMENT
Frank Harris and Ronald McCaffer
0 246 11370 7 paperback
176 pages Illustrated

MAPPING FROM AERIAL PHOTOGRAPHS
C. D. Burnside
0258 97035 9   320 pages Illustrated

HYDROGRAPHY FOR THE SURVEYOR AND ENGINEER
Alan E. Ingham
0 258 97165 7 paperback
148 pages Illustrated